GEOGRAPHIES OF EXCLUSION

Images of exclusion have characterized western cultures over long historical periods. In the developed society of racism, sexism and the marginalization of minority groups, exclusion has become the dominant factor in the creation of social and spatial boundaries.

Geographies of Exclusion seeks to identify the forms of social and spatial exclusion, and subsequently examine the fate of knowledge of space and society which has been produced by members of excluded groups. Evaluating writing on urban society by women and black writers, the author asks why such work is neglected by the academic establishment, suggesting that both the practices which result in the exclusion of minorities and those which result in the exclusion of knowledge have important implications for theory and method in human geography.

Drawing on a wide range of ideas from social anthropology, feminist theory, sociology, human geography and psychoanalysis, the book presents a fresh approach to geographical theory, highlighting the tendency of powerful groups to 'purify' space and to view minorities as defiled and polluting, and exploring the nature of 'difference' and the production of knowledge.

David Sibley teaches in the Department of Geography, University of Hull.

FOR AYANNA SIBLEY

GEOGRAPHIES OF EXCLUSION

Society and Difference in the West

DAVID SIBLEY

LONDON AND NEW YORK

First published 1995
by Routledge
2 Park Square, Milton Park, Abingdon, Oxon, OX14 4RN

Simultaneously published in the USA and Canada
by Routledge
270 Madison Ave, New York NY 10016

Reprinted 2001

Transferred to Digital Printing 2007

Routledge is an imprint of the Taylor & Francis Group

Typeset in Perpetua by
Solidus (Bristol) Limited

British Library Cataloguing in Publication Data
A catalogue record for this book is available from the British Library

Library of Congress Cataloguing in Publication Data
A catalogue record for this book has been requested

ISBN 0–415–11924–3
0–415–11925–1 (pbk)

Publisher's Note

The publisher has gone to great lengths to ensure the quality of this reprint but points out that some imperfections in the original may be apparent

Printed and bound by CPI Antony Rowe, Eastbourne

CONTENTS

PLATES

FIGURES

ACKNOWLEDGEMENTS

When I was sketching out some preliminary ideas for this book, I was encouraged by critical and constructive comments from Michael Dear and Peter Jackson. Peter, Stephen Frosh and Tristan Palmer also made useful suggestions about ways in which I could improve the final product. I have been fortunate in having opportunities to discuss my ideas on exclusion and boundary formation in seminars at St David's University College, Lampeter, Reading, Cambridge and Newcastle Universities, and the Women in Cities conference in Hamburg, 1992. I am grateful to a number of people who have contributed particular information and advice, including David Matless, James Sidaway, Jo Goodey, who provided some excellent photographs of children in urban environments, and Jacob Sibley, who was able to make Robert de Niro's mumblings in *Taxi Driver* intelligible. Midway through writing, exchanges with two psychologists, Geoff Lowe at Hull, and David Foxcroft at Portsmouth, with whom I worked on a project on teenage drinking, families and domestic space, contributed considerably to my appreciation of exclusionary processes in the home (and to different ways of looking at problems). Before and during the writing of this book, Chris Philo encouraged me to persist with my more bizarre ideas, some of which now appear rather conventional. I am very grateful to the technical staff at the School of Geography and Earth Resources, Hull University, particularly John Garner, Paul McSherry and Keith Scurr, for producing the photographs and diagrams. I would like to thank the following for permission to reproduce the following plates: Charlotte Hicks (4.2); Fleetway Editions Ltd (4.3); J. Walter Thompson Company Ltd (4.4); and the Fitzwilliam Museum, Cambridge (6.1).

INTRODUCTION

The human landscape can be read as a landscape of exclusion. This was clear to Engels in his observations on the industrial city, to Raymond Williams in his account of the landscapes of landed capital in eighteenth-century England in *The Country and the City*, and to Lewis Mumford, writing about Baroque cities in *The City in History*. Because power is expressed in the monopolization of space and the relegation of weaker groups in society to less desirable environments, any text on the social geography of advanced capitalism should be concerned with the question of exclusion. My purpose in writing this book, however, is not to provide a comprehensive account of exclusionary processes. There is already a substantial literature on the capitalist city which is, to some extent, concerned with exclusion, insofar as it is concerned with problems of access to urban resources, particularly housing, and associated spatial outcomes.[1] I would also leave off my agenda those programmes of exclusion which are starkly expressed in spatial terms and connect with clearly articulated ideologies, such as apartheid in South Africa and the 'race' policies of Nazi Germany, although I would not wish to suggest that these cases of oppression could not be further illuminated by geographical analysis.[2]

While this may seem like a perverse avoidance of central theoretical issues and crucial social and political problems, my intention in this book is to foreground the more opaque instances of exclusion, opaque, that is, from a mainstream or majority perspective, the ones which do not make the news or are taken for granted as part of the routine of daily life. These exclusionary practices are important because they are less noticed and so the ways in which control is exercised in society are concealed. One cue for my analysis comes from Paul Rabinow,[3] who has suggested that 'we need to anthropologize the West'. Rabinow argues that we need to 'show how exotic [the West's] constitution of reality has been; emphasize those domains most taken for granted as universal (this includes epistemology and economics); [and] make them seem as historically peculiar as possible'. To me this implies that we need to recognize as problems those aspects of life of which you might be unaware, particularly if you happen to be white, adult, male, and middle class, but which contribute to the oppression of others. Human geography, in particular, should be concerned with raising consciousness of the domination of space in its critique of the hegemonic culture. This has been the objective of Marxist analysis in human geography, but as a totalizing discourse Marxism has inevitably been insensitive to difference, almost as insensitive as the dominant capitalist culture which is the subject of Marxist critique. To get beyond the myths which secure capitalist hegemony, to expose oppressive practices, it is necessary to examine the assumptions about inclusion and exclusion which are implicit in the design of spaces and places. The simple questions we should be asking are: who are places for, whom do they exclude, and how are these prohibitions maintained in practice? Apart from examining legal systems and the practices of social control agencies, explanations of exclusion require an account of barriers, prohibitions and constraints on activities from the point of view of the excluded. I would agree with Jane Flax,[4] however, that there is no single oppressive reality, no single structure obscured by the images of the dominant culture, to uncover. She suggests that

> Perhaps reality can have 'a' structure only from the falsely universalizing perspective of the dominant group. That is, only to the extent that one person or group can dominate the whole will reality appear to be governed by one set of rules or to be constituted by one privileged set of social relations.

One part of the problem, then, is to identify forms of socio-spatial exclusion as they are experienced and articulated by the subject groups. These groups, however, may be seen as both dominant and subordinate, depending on the way in which they are categorized. Both men and women may experience exclusion as members of an oppressed minority group, for example, but men may be dominant in their relationship with women in a minority culture.[5] These different realities can be difficult to recognize, and the observer must appreciate that his or her own understanding of the socio-spatial worlds of others will inevitably be limited by his or her own background and perspectives. However, I still feel that it is possible, and certainly desirable, to represent other people's experience of socio-spatial exclusion while acknowledging that the question of positionality is one that has to be addressed.[6]

For the moment, rather than pursuing this argument in the abstract, I will comment on a few cases of exclusion which signal the specific kinds of socio-spatial issues which I will be considering in this book. The first concerns what is now a widely discussed problem, namely, the function of indoor shopping centres as social space.[7] These centres have become a significant mode of retail service provision in the developed capitalist economies, projected by both commercial and civic interests as progressive, and providing an improved environment for consumption and leisure for all the family. In the more extravagant developments, a fantasy world of imagined places is created, effectively removing consumption from associations with need. As Shields has observed in an account of the West Edmonton Mall in Canada, the model for several very large retailing developments in North America and Europe:

It fragments conventional geographical space and historical time with its wild combination of interior settings; evoking disparate times and places while it seeks to impose its own stable order on the ensemble. At the turn of a corner, one is in a simulated 'New Orleans'. Another corner – 'Paris'.

In comparable British developments, including the Meadowhall Shopping Centre, near Sheffield, which similarly recreates the romance of Paris and Florence under one roof, and the Metro Centre in Gateshead, their exoticism has stimulated a new form of holiday experience. For some, a coach trip to the shopping centre has become a substitute for a day by the sea, in Blackpool, Scarborough or Skegness. Such places clearly do provide an attractive alternative to a traditional shopping street, polluted by vehicle exhausts and exposed to the weather (and they may be rather more appealing than a decaying seaside resort). Thus, a British television documentary on the Metro Centre in Gateshead focused in a positive way on the characteristic features of international consumption style and the consumers, all apparently white, middle-class nuclear families, the kind of public which populates architects' sketches. The documentary had a rather different sub-text, however. Out of sight in the control room, employees of the private security firm which polices the centre had their eyes fixed on closed-circuit television screens. They were looking for 'undesirables', mostly groups of teenage boys who did not fit the family image projected by the company. When they were located, security guards evicted them, not just from the building but from the precinct. Such actions point to the fact that shopping centres like this one constitute a kind of ambiguous, seemingly public but actually private space. There are implicit rules of inclusion and exclusion in a built form that contribute to the structuring of society and space in a way which some will find oppressive and others appealing. 'Being in the tightly policed, semi-private interior of a mall is quite different from being "on the street". "No loitering", as the signs in the mall say. Certain types of comportment are expected'.[8] In the shopping centre management's response to the presence of adolescents, maybe not consuming very much, in a place dedicated to consumption by the family, there is a connection between the function and design of the space as determined by commercial interests and design professionals, architects and planners, and the construction of one group of the population as 'deviant', out of place, and threatening the projected image of the development. Again, Shields notes that the shopping mall introduces 'an unheard of degree of surveillance, with almost Orwellian overtones, into daily life', and, in this controlled environment, teenagers who have few other places in which to congregate are one of the principal groups targeted by the security guards. Their presence necessarily constitutes deviance.[9] Comparing this with my own experience as a teenager in the 1950s, sitting for hours over a cup of coffee in an ABC café in a north London suburb, undisturbed by staff, it appears that the boundaries between the consuming and non-consuming public are strengthening, with non-consumption being constructed as a form of deviance at the same time as spaces of consumption eliminate public spaces in city centres.

This view gains some empirical support in a number of studies of teenage sub-cultures. For example, in a Home Office study of 'downtown drinkers' in the planned shopping precinct in Coventry,[10] the writer reported that:

> Unruly groups of young people were seen as a problem by approximately two out of three interviewees. As with litter, the problem is not just a local matter. Throughout the country, shopping centres often serve as convenient places for youth subcultures to meet – places to which there may be a lack of obvious alternatives. On the other hand, *the mere sight of such groups*, however rarely they actually infringe any laws, can be alarming to others. This is a delicate issue. [Attempts] to exclude youth groups from shopping centres are likely to bring further problems, and may also be quite unjust. Nevertheless, in Coventry, perception of unruly groups of young people as a common problem *was significantly associated with the fear of crime*. This link was stronger than that between perceptions of litter and fear of crime but not as powerful as that between perceptions of public drinkers and fear of crime [my italics].

This quotation, like the television documentary on the shopping centre, suggests that it is not adolescent males as a social category, or even 'unruly' groups of young people, *per se*, who are seen as threatening; rather, it is their presence in spaces which comprise part of 'normal family space' which renders them discrepant and threatening. Exclusion may be an unintended consequence of commercial development. Adolescents will be acutely aware of discrimination against them, while their exclusion is much less likely to impinge on the consciousness of conforming adults.

In the interaction of people and the built environment, it is a truism that space is contested but relatively trivial conflicts can provide clues about power relations and the role of space in social control. This is demonstrated in a second example, dredged from memory, which concerns an incident in Philadelphia in the late 1960s. At this time, hippies were still a threatening species in Philadelphia, the category 'hippy' embracing just about any man with long hair or woman wearing beads. Rittenhouse Square, in the city centre, was popular with slightly non-conforming people at the weekend; in it was a low wall which was a convenient place to sit. One warm Sunday afternoon, there were a lot of people sitting on the wall, some playing acoustic guitars, but mostly just chatting and enjoying the sunshine. At some point, a park guard started to order people off the wall on the grounds that it was *not* a place to sit. The wall, he asserted, was there to separate the path from the grass. It was definitely not to sit on. Almost everyone acquiesced. This might have been because the park guard, who, like many agents of social control in the United States, was equipped with a revolver and a night-stick, appeared intimidating. It could also be the case that this group of middle-class American youth, having been brought up in conformist communities, were accustomed to accepting authority despite their trappings of non-conformity.

There are two aspects of this incident which are of more general significance. The first concerns ambiguity. To the park guard, the function of the wall was unambiguous. It was simply a boundary between one kind of space and another and, apparently, he could not conceive of alternative interpretations. His job was to police the wall, to ensure its sanctity and prevent its violation. It may be reading too much into the incident, but his behaviour appeared to fit a pattern noted in a number of studies of the authoritarian personality, following Adorno's early study of the psychology of authoritarianism.[11] Shils suggested that authoritarians were distrustful and suspicious, that they had *an intolerance of ambiguity*, and, thus, differentiated clearly between those on the outside, the 'other', and the relevant in-group.[12] Similarly, Rokeach[13] suggested that authoritarian individuals '[protected] inner weaknesses by a ready acceptance of the views of higher authorities and by forming unambiguous judgements which rigidly separate, into distinct categories, objects of approval and those of disapproval'.

The issue is not just about an unsuspecting park guard overloaded with theoretical meaning. Apart from the park guard's own perception of non-conformity, the social status of Rittenhouse Square also contributed to the representation of its hippy-ish occupants as conspicuously deviant. The square was surrounded by solid apartment buildings occupied by affluent middle-aged and elderly residents who saw the hippies as polluting 'their' space. In fact, the park guard admitted to me that he had been told to clear the square of young people because their presence offended the residents. The arbitrary use of power by the guard thus reflected a more fundamental aspect of power relations. The square as a contested public space exposed the conflictual nature of social relations and the design of the square itself assumed symbolic importance in this conflict. It should not be seen just as an arena where this particular power game was played, however, but as one instance of the interaction of space and people which forms part of the routines for the reproduction of power relations in an advanced capitalist society.[14]

The policing of Rittenhouse Square, a rather unsubtle example of social control, might be compared with many instances of exclusion where boundaries are drawn discretely between dominant and subordinate groups. Martin Walker notes the spread of the private pool club in the United States, an institution, like the whites-only golf club, which continues 'the discrete and self-deceiving way of modern American apartheid. It is now justified as a way to avoid the crowds, crime and drugs of the municipal pools, these being code words which are used to signify black people.'[15] Elsewhere, Mike Davis has captured the helplessness of the poor and homeless in the large North American city, faced with exclusionary developments by corporate capital.[16] Talking to a black, homeless man in downtown Los Angeles, Davis comments: 'In front of us, tens of thousands of poor people, homeless people; at back opulence, affluence, Bunker Hill, the new L.A.' He then asks: 'Could you walk up there?' and the man replies: 'If they were to catch me in that building, they would have so much security on my ass, I would probably be in jail in five minutes.' Again exclusion is felt acutely, but the homeless are rendered invisible to the affluent downtown workers by the spatial separations of city centre development which keep the underclass at a distance.

These examples give some indication of the concerns of this book, exclusions in social space which may be unnoticed features of urban life. It is the fact that exclusions take place routinely, without most people noticing, which is a particularly important aspect of the problem. In an attempt to make these practices more transparent, what I try to do in the first part of the text is to define attitudes to others which inform exclusionary practices and to set the control problem in the broader context of the cultures of modern societies.

I then try to show how the processes of control are manifested in the exclusion of those people who are judged to be deviant, imperfect or marginal. A study of exclusion, however, is necessarily concerned with inclusion, with the 'normal' as much as the 'deviant', the 'same' as well as the 'other', and with the credentials required to gain entry to the dominant groups in society. Thus, I focus on processes of boundary erection by groups in society who consider themselves to be normal or mainstream. The curious practices of this majority, the oddness of the ordinary which is examined microscopically by authors and playwrights from Jane Austen to Mike Leigh, have been neglected in social geography, and one of the purposes of this book is to rectify this omission.

My treatment of space and society is concerned particularly with symbol, ritual and myth, taking cues from social anthropology and psychoanalysis, subjects which have not been overly concerned with space but which provide many useful analogues for spatial problems. I would argue that many social problems can be profitably spatialized, but, at the same time, a human geography which attempts to assume a distinct identity within social science is necessarily impoverished. For example, it seems to me that the concern of social anthropology with representation, imagery and alternative world-views should also be central to human geography, hence 'geographies' in the title of the book. To uncover these diverse geographies, reflecting varied experiences and interpretations of space and place, involves drawing on a wide range of ideas located elsewhere in the social sciences and the humanities. A post-disciplinary perspective on social and spatial problems is preferable to viewing the world from within conventional subject boundaries.

In Part I, I first attempt to build up pictures of the rejecting and the rejected and then travel along several theoretical avenues in order to identify exclusionary processes affecting both groups and individuals. In addition to theories of socio-spatial structuring, this section makes reference to psychoanalytic theories of the self, which assumed greater importance as the writing progressed. This was partly because I was trying to familiarize myself with this literature while working on the text but also because some ideas from psychoanalysis seemed to connect with what were, for me, more familiar arguments about boundary formation developed in social anthropology and human geography. I would not claim that this account achieves any real synthesis, but it does suggest some connections between individual and group behaviour, and between environment and behaviour, which might be integral elements of the problem.

These theoretical arguments connect with instances of exclusion at different spatial scales, starting with the home and moving out to the nation-state and questions of geopolitics. Exclusion in the home, in the locality and at the national level are not discrete issues. A number of reciprocal relationships are examined and there is, inevitably, considerable cross-referencing in this part of the book. While there are common strands to the argument here, the problems considered are very different, ranging from conflicts within families and homes to international relations.

In Part II, I get away from the usual subjects of geographical analysis to consider academics as subjects, but what I claim in this section is that we can use the same arguments to explain the exclusion of knowledge as to explain the exclusion of discrepant others. I suggest that the production of knowledge involves both the exclusion of knowledge which is deemed dangerous and the exclusion of some categories of intellectual. The processes of social segregation observable in the modern city, for example, are mirrored in the segregation of knowledge producers. The defence of social space has its counterpart in the defence of regions of knowledge. This means that what constitutes knowledge, that is, those ideas which gain currency through books and periodicals, is conditioned by power relations which determine the boundaries of 'knowledge' and exclude dangerous or threatening ideas and authors. It follows that any prescriptions for a better integrated and more egalitarian society must also include proposals for change in the way academic knowledge is produced.

I do not attempt in this book to give an account of exclusion in advanced capitalist societies which covers all salient forms of difference. There would be a long list of these, including exclusion based on race, gender, sexuality, age, and mental and physical disability. What I hope to do, however, is to clarify some of the spatial and social boundary processes which separate some groups and individuals from society and render deviant those who are different. At the same time, I suggest that social scientists need to look more closely at their own practices and develop critiques of their work which parallel their analyses of the marginalized and oppressed.

NOTES

1. Thus, much of David Harvey's work could be read as a (class-based) geography of exclusion. His essay 'Class structure and residential differentiation' (in *The Urban Experience*, Basil Blackwell, Oxford, 1989, pp. 109–124) is specifically concerned with closure and exclusion as they operate in the property market. In a similar theoretical vein, we could note Blair Badcock's

Unfairly Structured Cities, Basil Blackwell, Oxford, 1984. Weberian closure theory is also concerned with exclusion – through the erection of barriers to entry into more privileged groups. Closure theory has been given a spatial dimension, particularly in Christopher Husbands's work on racism. See his 'East End racism, 1900–1980', *The London Journal*, 8, 1982, 3–26.

2. Rössler's study of the connection between central place theory and the concept of *lebensraum* in Nazi Germany demonstrates that there is considerable scope for the kind of research on fascist and other authoritarian regimes which explores ideology, spatial theories and spatial practices (Mechtild Rössler, 'Applied geography and area research in Nazi society: central place theory and planning, 1933 to 1945', *Environment and Planning D: Society and Space*, 7 (4), 1989, 363–400).

3. Paul Rabinow, 'Representations are social facts: modernity and post-modernity in anthropology', in James Clifford and George Marcus (eds), *Writing Culture*, University of California Press, Berkeley, 1986, 234–261. This echoes Robert Park's recommendation in his 1925 essay on the city:

> Anthropology, the science of man, has been mainly concerned up to the present with the study of primitive peoples. But civilized man is quite as interesting an object of investigation and, at the same time, his life is more open to observation and study. Urban life and culture are more varied, subtle and complicated, but the fundamental motives in both instances are the same. The same patient methods of observation which anthropologists like Boas and Lowie have expended on the study of the life and manners of the North American Indian might be even more fruitfully employed in the investigation of the customs, beliefs, social practices and general conceptions of life prevalent in Little Italy or the lower North Side in Chicago or in recording the more sophisticated folkways of the inhabitants of Greenwich Village and the neighborhood of Washington Square, New York.
>
> (Robert Park, *The City*, University of Chicago Press, Chicago, 1925, p. 2)

4. Cited by Linda Nicholson (Linda Nicholson (ed.), *Feminism/Postmodernism*, Routledge, London, 1990, p. 6).

5. In patriarchal Gypsy communities, for example, women suffer double exclusion, as women and as members of a marginalized minority.

6. However, agonizing over position leads to authors denying the possibility of writing with any authority about anybody other than their own social group, which may be quite narrowly defined. Given appropriate methods of investigation, I feel that some generalization about those with different world-views is possible and desirable, although there is always a risk of distortion and misrepresentation which can only be guarded against by repeated engagement with other groups.

7. Rob Shields's 1989 essay is one of the more thoughtful studies of shopping centres, but David Harvey also makes some relevant comments on the blurring of public and private space in the new arenas of consumption. See Rob Shields, 'Social spatialization and the built environment: the West Edmonton Mall', *Environment and Planning D: Society and Space*, 7, 1989, 147–164, and David Harvey, 'Postmodern morality plays', *Antipode*, 24, 1992, 300–326.

8. Shields, op. cit.

9. With a thorough application of surveillance technology, the shopping centre can become a 'panoptican mall', echoing Jeremy Bentham's design for a model prison. See Mike Davis, *City of Quartz*, Verso Press, London, 1990, pp. 240–244, on 'the mall-as-panoptican-prison'

as it has been realized in inner-city Los Angeles.

10. M. Ramsay, *Downtown Drinkers: The perceptions and fears of the public in a city centre*, Crime Prevention Unit, Paper 19, Home Office, London, 1989.

11. Theodor Adorno, *et al.*, *The Authoritarian Personality*, Norton, New York, 1982.

12. Edward Shils, 'Authoritarianism: "right" and "left"', in R. Christie and M. Jahoda (eds), *Studies in the Scope and Method of the Authoritarian Personality*, Free Press, Glencoe, 1954.

13. Michael Rokeach, *The Open and Closed Mind*, Basic Books, New York, 1960.

14. David Harvey makes similar points about social relations in Tompkins Square Park, New York City, where 'On a good day, we could celebrate the scene within the park as a superb example of urban tolerance for difference', but 'on a bad day. . . so-called forces of law and order battle to evict the homeless, erect barriers between violently clashing factions. The park then becomes a locus of exploitation and oppression' (1992, op. cit.).

15. Martin Walker, *Guardian*, 26 May 1990.

16. *Rear Window*, Channel 4 TV, London, 1991. As Davis puts it, rather floridly: 'The Downtown hyperstructure – like some Buckminster Fuller post-Holocaust fantasy – is programmed to ensure a seamless continuum of middle-class work, consumption and recreation, without unwonted exposure to Downtown's working-class street environments' (op. cit., 1990, p. 231).

Part I

GEOGRAPHIES OF EXCLUSION

1

FEELINGS ABOUT DIFFERENCE

> The senior partner of a well-known professional firm around here put his home on the market with us and he said: 'You sent me a Mr Shah and you sent me a Mr Patel and you sent me a Mr Whatever-it-was.' He said: 'I recognize that a lot of the big money comes from several thousand miles east of Dover nowadays, and I don't want you to think that I've got any prejudice at all, but would you be able to send me an Englishman one day?'
>
> (Suburban London estate agent)[1]

There are several possible routes into the problem of social and spatial exclusion. I want to start by considering people's feelings about others because of the importance of feelings in their effect on social interaction, particularly in instances of racism and related forms of oppression. If, for example, we consider the question of residential segregation, which is one of the most widely investigated issues in urban geography, it could be argued that the resistance to a different sort of person moving into a neighbourhood stems from feelings of anxiety, nervousness or fear. Who is felt to belong and not to belong contributes in an important way to the shaping of social space. It is often the case that this kind of hostility to others is articulated as a concern about property values but certain kinds of difference, as they are culturally constructed, trigger anxieties and a wish on the part of those who feel threatened to distance themselves from others. This may, of course, have economic consequences.

Feelings about others, people marked as different, may also be associated with places. Nervousness about walking down a street in a district which has been labelled as dangerous, nauseousness associated with particular smells or, conversely, excitement, exhilaration or a feeling of calm may be the kinds of sensations engendered by other environments. Repulsion and desire, fear and attraction, attach both to people and to places in complex ways. Central to this question is the construction of the self, the way in which individual identity relates to social, cultural and spatial contexts. In this chapter, I will suggest some of the connections between the self and material and social worlds, moving towards a conception of the 'ecological self'.[2]

ALTERNATIVE PERSPECTIVES ON THE SELF

Central to early visions of the self was the idea of human individuality.[3] Rationalist philosophers recognized that only human beings were consciously aware of their own life, which gave them the capacity to act autonomously. Nineteenth-century romanticism similarly encouraged a view of the free spirit, and this notion of the self was reinforced by capitalist forms of social organization according to which people are highly individuated and assumed to have control of their own destinies. The subject was thus detached from his or her social milieu.

A shift in conceptions of the self was signalled by Freudian psychoanalysis. Freud situated the self in society and argued for connections between the developing self and the material world. Central to his thesis was the unconscious, that 'aspect of psychoanalysis that directly challenges the emphasis in Western thought on the power of reason and rationality, of reflective and conscious control over the self'. Although Freud suggested that on one level the unconscious was detached from reality, on another level 'it is deeply entwined with the needs of the human body, the nature of external reality, and actual social relations'.[4] The importance of external reality for the psyche was outlined in *Civilization and its Discontents*, published in 1929. In this book, Freud wrote about the repression of libidinal desires specifically in relation to the materialism of capitalist societies. He claimed that one form of repression was an excessive concern with cleanliness and order.[5] Personal hygiene, for example, is widely accepted as desirable on medical and social grounds, but it removes bodily smell as a source of sexual stimulation. Washing and deodorizing the body has assumed a ritual quality and in some people can become obsessive and compulsive. This kind of observation raises issues about the role of dominant social and political structures in the sublimation of desire and the shaping of the self. What are the sanctions against a group or an individual represented as dirty or disorderly? In *Civilization and its Discontents* Freud brackets cleanliness and order, both distancing the subject from the uncertainties and fears of the urban-industrial environment. However, as Smith observes, 'order is a part of the tragedy of modern urban culture: it brings frustration but it cannot be done without'.[6]

OBJECT RELATIONS THEORY

Freud's psychoanalytical writing provides a starting point for an examination of relationships between the self and the social and material world. This is the field of object relations theory, which, for Freud, referred to the infant's relationship to the humans in its world, but it is a theory which has been generalized to include non-human aspects of the object world, a wider environment of human and material objects, and extended beyond infancy. The latter is a particularly important contribution of Erik Erikson,[7] who argued that 'the personality is engaged with the hazards of existence continuously, even as the body's metabolism copes with decay'. Erikson tried to model the changes in the self over the life course in the form of eight stages of ego development which he described as the 'Eight Ages of Man'. The details of his schema are not as important as the idea of change throughout life, resulting from continuous engagement with the environment.

Object relations theory, as it has been reworked by psychoanalytical theorists since Freud, has an important role in my argument.

> Object relations theory assumes that from birth, the infant engages in formative relations with 'objects' – entities perceived as separate from the self, either whole persons or parts of the body, either existing in the external world or internalized as mental representations.[8]

It suggests ways in which boundaries emerge, separating the 'good' and the 'bad', the stereotypical representations of others which inform social practices of exclusion and inclusion but which, at the same time, define the self. In the following account I draw primarily on Melanie Klein's work but I also refer to authors who put more emphasis on the social context of psychoanalytic theory, particularly Julia Kristeva and Constance Perin.[9]

Klein,[10] like Freud, focuses on infancy but she provides a clear and quite convincing argument about the development of the social self. Her argument is that in the process of birth and immediately after birth the infant experiences anxieties associated with the initial discomforts of being – light, cold, noise, and so on, but comfort comes from being held by the mother and from breast-feeding, which make possible 'the infant's first loving relation to a person [object]'. The infant experiences a feeling of one-ness with the mother, who, 'in the first few months . . . represents to the child the whole of the external world'. Necessarily, both good and bad emanate from the mother because she is the source of all of the earliest experience of social relationships. The mother is, therefore, both a good and a bad object. However, this initial, pre-Oedipal one-ness with the mother is lost as the child develops a sense of border, a sense of self-hood, and a sense of the social. This comes about through a combination of two processes. The first is introjection, whereby 'the situations the infant lives through and the objects he or she encounters are taken into the self and become a part of inner life. Inner life cannot be evaluated . . . without these additions to the personality that come from continuous introjection.' Klein identifies a simultaneous process of projection, 'which implies that there is a capacity of the child to attribute to other people . . . feelings of various kinds, predominantly love and hate.' However, objects are not necessarily either polarized or in balance. As she recognized, if projection is predominantly hostile, empathy with others is impaired. Conversely, if the child 'loses itself entirely in others', a condition of excessive introjection, it becomes incapable of independent, objective judgement. Seeing the world exclusively as 'good' or 'bad' is, in Klein's terms, the paranoid-schizoid position. Most personalities exhibit finer gradations of goodness and badness, however, and do, in her view, achieve a kind of balance.

This concern with balance as a desirable personality characteristic is found in much psychotherapy and it represents a more optimistic view of the self than that of those social philosophers who see the psyche as buffeted by social forces.[11] However, this is not the main issue here. Whatever the contribution of introjection and projection to the shape of the emerging self, the significant contribution of Klein is her view of the emerging sense of border, of separateness and self, as a social and cultural process. As Hoggett interprets Klein's argument,[12] this sense of border is confirmed through the construction of 'an endless series of misrepresentations, all of which share an essential quality, the quality of otherness, of being not-me'.

Separation from the mother has a cost, that is, an anxiety which results from a fear of merging again with the mother. This internal fear is externalized, and security is gained through associating fear with an external threat. The threat comes from an array of 'others' which provide protection for the self. They define the boundary which protects against dissolution. Countering this fear of dissolution, however, merging is associated with comfort and pleasure and separateness with loneliness. This suggests that some people will have a greater boundary consciousness than others. Some will embrace difference, gain pleasure and satisfaction from merging, while others will reject difference. Most personalities will have a mix of both characteristics.

The self is a cultural production. The perpetual restructuring of the self takes place through what Lacan calls the 'symbolic order', which includes social and cultural symbolism. The sense of border which emerges in infancy is not an innate sense but a consequence of relating to others and becoming a part of a culture. Thus, the boundary between the inner (pure) self and the outer (defiled) self, which is initially manifest in a distaste for bodily residues but then assumes a much wider cultural significance, derives from parents and other adults who are, by definition, socialized and acculturated. However, some writers have claimed that the marking of this boundary by bodily waste is particularly characteristic of modern western cultures.[13] Associations are made between faeces, dirt, soil, ugliness and imperfection, but these are particularly puritanical, western obsessions. This initial sense of border in the infant in western societies becomes the basis for distancing from 'others', but the question of otherness can only be discussed meaningfully in a cultural context, for example, in relation to racism or to a 'colonial other'. Thus, enthusiasm for psychoanalysis in the social sciences has to be tempered with an acknowledgement of its failure to deal adequately with difference.

An essentialist notion of the bounded self has been challenged in another way by Constance Perin.[14] She argues that the fear of mixing and merging, which is expressed in the imperative of distancing from shit, reflects a particularly masculine concern for autonomy and separateness. It is the mother who experiences one-ness with the baby: she is the one who is primarily responsible for nappy changing and who has the greatest familiarity with the infant's bodily residues. Thus, female and male roles in child rearing reflect the masculine value of autonomy and the feminine value of merging and a tolerance for difference. This is, in itself, a polarized and essentialist argument, but it does serve to demonstrate how culture might affect conceptions of the boundary between self and other.

ABJECTION

Maintaining the purity of the self, defending the boundaries of the inner body, can be seen as a never-ending battle against residues – excrement, dead skin, sweat, and so on, and it is a battle that has wider existential significance. To quote Julia Kristeva: 'Excrement and its equivalents (decay, infection, disease, corpse [*sic*], etc) stand for the danger to identity that comes from without: the ego threatened by the non-ego, society threatened by its outside, life by death'.[15] Kristeva's reflections on the pure and defiled in her essay on abjection seem particularly compelling. She maintains that the impure can never be completely removed: 'We may call it a border; abjection is above all ambiguity. Because, while releasing a hold, it does not radically cut off the subject from what threatens it – on the contrary, abjection acknowledges it to be in perpetual danger'.[16]

What Kristeva describes as abject is 'opposed to *I*', it is 'radically excluded', but it is always a presence. She follows Georges Bataille, who defines abjection as 'merely the inability to assume with sufficient strength the imperative act of excluding abject things'.[17] Her view of the abject as some thing, always there, '[hovering] at the borders of the subject's identity, threatening apparent unities and stabilities with disruption and possible dissolution', as Elizabeth Gross puts it,[18] points to the importance of anxiety, a desire to expel or to distance from the abject other as a condition of existence. This hovering presence of the abject gives it significance in defining relationships to others. It registers in nervousness about other cultures or about things out of place. In another attempt to define the abject, Gross claims that

> [It] is an impossible object, still a part of the subject: an object the subject strives to expel but which is ineliminable. In ingesting objects into itself or expelling objects from itself, the 'subject' can never be distinct from these 'objects'. The ingested objects are neither part of the body nor separate from it.[19]

Yet, the urge to make separations, between clean and dirty, ordered and disordered, 'us' and 'them', that is, to expel the abject, is encouraged in western cultures, creating feelings of anxiety because such separations can never be finally achieved. This anxiety, as I hope to demonstrate in the next two chapters, is reinforced by the culture of consumption in western societies. The success of capitalism depends on it.

Feelings about others on one level register as *sensations* associated with the abject – people, things and places in various combinations. Kristeva describes a sense of the abject in visceral terms. Food loathing, for example,

> is perhaps the most elementary and most archaic form of abjection. When the eyes see or the lips touch that skin on the surface of milk – harmless, thin as a sheet of cigarette paper, pitiful as a nail paring – I experience a gagging sensation and, still farther down, spasms in the stomach, the belly.

Such sensations can become a part of social experience, however, as Alain Corbin suggests in his olfactory tour through French culture.[20] Historically, Corbin suggests that the bourgeois self separated itself from the working-class other through smell or a fear of smell, which returned the bourgeois to the original source of abjection, defilement associated with bodily residues. Similar feelings of abjection, I will suggest, attach also to place, but to understand the connection between abject things, people and places requires an appreciation of 'the generalized other'.

THE GENERALIZED OTHER

The concept of 'the generalized other' provides a means of spatializing the problem and producing what we might describe as an ecological account of the self, one which situates the self in a full social and cultural context. The term 'generalized other' was first used by George Herbert Mead, who noted the elision of people and objects to whom the child relates in developing a sense of self. He argued that:

> It is possible for inanimate objects, no less than for other human organisms, to form parts of the generalized and organized – the completely socialized – other for any given individual . . . Any thing – any object or set of objects, whether animate or inanimate, human or animal, or merely physical – towards which he acts, or to which he responds socially, is an element in what for him is the generalized other.[21]

Mead's interpretation of the relationship between self and other has fundamental implications for geographical studies of social interaction because it locates the individual in the social and material world. Ian Burkitt gives prominence to Mead's object relations theory for similar reasons.[22] He argues that:

> Mead's conception of the self and the psychical apparatus is more useful than Derrida's or Freud's in studying *the body in action*. That is because Mead recognized the practical nature of

the psyche, *that it is always connected to social practice* and does not exist in some separate textual or mental domain. Whereas Derrida and Freud struggled with the metaphor of the mystic writing-pad [for Derrida, a cultural and historical text written into the unconscious, positioning the subject in a textual world], Mead conceptualized that which remains open to new experiences and information as the active person *in their various social locations and settings* (the 'I'). It is the embeddedness in social contexts that allows the individual to be constantly receptive to new stimuli, while at the same time the body carries the forms of history in terms of the cultural image of the self and the disciplines involved in social interaction (the 'me'). So the 'I' and the 'me' are not just psychical but also bodily [my italics].

The social positioning of the self means that the boundary between self and other is formed through a series of cultural representations of people and things which frequently elide so that the non-human world also provides a context for selfhood. To give one example of this kind of cultural representation, in racist discourse animals represented as transgressive and therefore threatening unsullied categories of things and social groups, like rats which come out of the sewers and spread disease, have in turn been used to represent threatening minority groups, like Jews and Gypsies, who are thus constructed as bad objects to which the self relates. To animalize or de-humanize a minority group in this way, of course, legitimates persecution. Interestingly from a geographical perspective, one of the few applications of Mead's generalized other has been in studies of the organization of domestic space by Csikszentmihalyi and Rochberg-Halton,[23] where things in the home which are both positively valued and rejected are seen to have a defining role in relation to the self. They note, in particular, that

the impact of inanimate objects in this self-awareness process is much more important than one would infer from its neglect. Things tell us who we are, not in words but by embodying our intentions. In our everyday traffic of existence, we can also learn about ourselves from objects, almost as much as from people.[24]

People and things come to stand for each other, Csikszentmihalyi and Rochberg-Halton suggest, so that object relations can include relating to others through the material environment. Thus, for one woman,

her home environment reflects an expanded boundary of the self, one that includes a number of past and present relationships. The meanings of the objects she is surrounded by are signs of her ties to this larger system of which she is a part.[25]

This seemed to me to be a promising but little-developed direction for research, one in which the signing of spaces could be examined specifically in relation to the social self. It had considerable implications for studies of inclusion and exclusion in other spaces.

CONCLUSION

Object relations theory has a particular value for this study, with its focus on the discrepant. The 'other' could be examined solely as a social category, but feelings about others, the ambivalent sensations of desire and disgust which energize interpersonal and social relations, require an understanding of the self. The emphasis in Kleinian psychoanalysis in particular on introjection and projection connects the self and society, and this then leads to questions about the nature of the border separating self and other as it is constructed in different cultures. The context for this study is 'the West' or 'capitalist society', which are, admittedly, heterogeneous categories, but we can probably talk meaningfully about the characteristic anxieties of the western self, which are explored, for example, by Kristeva in her account of the abject. Abjection seems to me the key to an understanding of exclusion, although the social and spatial contexts of abjection need considerable elaboration.

In subsequent chapters, the geographies of exclusion, the literal mappings of power relations and rejection, are informed by the generalized other. Apart from the collapse of categories like the public and the private which I see as a necessary feature of these geographies, the generalized other of object relations theory gives an invitation to open up debates about otherness, to examine the interconnections of people and things as they constitute and are constituted by places, what I take to be the ecological self (and the ecological other). This has to be taken one step at a time, however. I first look at social boundaries, filling in some details about the people who erect the boundaries and those who are excluded by them, and I then consider the issue of exclusionary landscapes as they have developed in different times and places.

NOTES

1. Daniel Meadows, *Nattering in Paradise: A word from the suburbs*, Simon and Schuster, London, 1988, p. 40.

2. This term comes from Ulric Neisser, 'Five kinds of self-knowledge', *Philosophical*

Psychology, 1 (1), 1988, 35–59. Neisser makes a number of interesting points about the ways in which the self relates to the environment, although he does not say what the environment is. He suggests (1) that we perceive ourselves as embedded within the environment, and acting with respect to it; (2) that the self and the environment exist objectively; information about the self allows us to perceive not only the location of the ecological self but also the nature of its interaction with the environment; (3) that much of the relevant information is kinetic, i.e. relating to movement. Optical structure is particularly important, but self-specifying information is often available to several perceptual modalities at once; and (4) the ecological self is veridically perceived from earliest infancy, but self-perception develops with increasing age and skill. I use the term 'ecological self' in a more inclusive sense, to refer to the self defined in relation to people, things and places, as they relate to each other.

3. Ian Burkitt provides an excellent account of the western self in an historical context in I. Burkitt, 'The shifting concept of the self', *History of the Human Sciences*, 7 (2), 1994, 7–28.

4. Anthony Elliott, *Social Theory and Psychoanalysis in Transition*, Basil Blackwell, Oxford, 1992, pp. 16–17.

5. Michael Smith, *The City and Social Theory*, Basil Blackwell, Oxford, 1980, pp. 57–58.

6. ibid., p. 58.

7. Erik Erikson, *Childhood and Society*, Penguin, Harmondsworth, 1970.

8. Claire Kahane, 'Object relations theory', in Elizabeth Wright (ed.), *Feminism and Psychoanalysis: A critical dictionary*, Basil Blackwell, Oxford, 1992, p. 284.

9. Constance Perin is a social anthropologist who has drawn on psychoanalytical theory in her studies of mainstream America. Julia Kristeva's writing is impossible to categorize, but her psychoanalytical reflections connect with a number of social and political issues. I find her writing provocative and stimulating. Her arguments are wide ranging, touching on religion, literature, social anthropology and the politics of difference.

10. Melanie Klein, *Our Adult World and its Roots in Infancy*, Tavistock Pamphlet 2, London, 1960.

11. Notably, Norbert Elias: 'By his birth [the subject] is inserted into a functional complex with a quite definite structure; he must conform to it, shape himself in accordance with it and perhaps develop further on its basis' (N. Elias, *The Society of Individuals*, Basil Blackwell, Oxford, 1991, p. 14).

12. Paul Hoggett, 'A place for experience: a psychoanalytic perspective on boundary, identity and culture', *Environment and Planning D: Society and Space*, 10, 1992, 345–356.

13. Constance Perin asserts that 'Evil is embodied according to Western beliefs in excrement: Defilement, Devilry, Disease and Sin shape this conceptual system' (C. Perin, *Belonging in America*, University of Wisconsin Press, Madison, 1988, p. 178).

14. ibid., pp. 198–207.

15. Julia Kristeva, *Powers of Horror*, Columbia University Press, New York, 1982, p. 71.

16. ibid., p. 3.

17. ibid., p. 56.

18. Elizabeth Gross, 'The body of signification', in J. Fletcher and A. Benjamin (eds), *Abjection, Melancholia and Love: The work of Julia Kristeva*, Routledge, London, 1990, pp. 80–103.

19. Elizabeth Gross, 'Julia Kristeva', in Elizabeth Wright (ed.), *Feminism and Psychoanalysis: A critical dictionary*, Basil Blackwell, Oxford, 1992, p. 198.

20. Alain Corbin, *The Fragrant and the Foul: Odor and the French social imagination*, Harvard iiversity Press, Cambridge, Mass., 1986.

21. George Herbert Mead, *Mind, Self and Society*, Chicago University Press, Chicago, 1934.

22. Burkitt, op. cit., p. 23.

23. Mihalyi Csikszentmihalyi and Eugene Rochberg-Halton, *The Meaning of Things: Domestic nbols and the self*, Cambridge University Press, Cambridge, 1981.

24. ibid., p. 91.

25. ibid., p. 104. In a later essay, Eugene Rochberg-Halton suggested that

: meaning of things one values is not limited just to the individual object itself but also includes the spatial ntext in which the object is placed, forming a domain of personal territoriality. In other words, the :kground context or gestalt of the thing also reveals something and results show how different rooms 'eal different conceptions of the self

(Eugene Rochberg-Halton, 'Object relations, role models and the cultivation of the self', *Environment and Behavior*, 16 (3), 1984, 335–368)

2

IMAGES OF DIFFERENCE

The determination of a border between the inside and the outside according to 'the simple logic of excluding filth', as Kristeva puts it, or the imperative of 'distancing from disgust' (Constance Perin) translates into several different corporeal or social images which signal imperfection or a low ranking in a hierarchy of being. Exclusionary discourse draws particularly on colour, disease, animals, sexuality and nature, but they all come back to the idea of dirt as a signifier of imperfection and inferiority, the reference point being the white, often male, physically and mentally able person. In this chapter, I will discuss ways in which psychoanalytical theory has been used in the deconstruction of stereotypes, those 'others' from which the subject is distanced, and I will then examine some of the particular cultural sources of stereotyping in western societies. Stereotypes play an important part in the configuration of social space because of the importance of distanciation in the behaviour of social groups, that is, distancing from others who are represented negatively, and because of the way in which group images and place images combine to create landscapes of exclusion. The issues I examine concern oppression and denial. I try to show how difference is harnessed in the exercise of power and the subordination of minorities.[1]

STEREOTYPES

The reception and acceptance of stereotypes, 'images of things we fear and glorify', as Sander Gilman puts it,[2] is a necessary part of coming to terms with the world. In the following passage from his psychoanalytical account of the deep structure of stereotypes, Gilman assigns a central role to stereotyping in the structuring or bounding of the self:

The child's sense of self splits into a 'good' self which, as the self mirroring the earlier stage of the complete control of the world [the stage of pre-Oedipal unity with the mother] is free from anxiety, and the 'bad' self which is unable to control the environment and is thus exposed to anxieties. The split is but a single stage in the development of the normal personality. In it lies, however, the root of all stereotypical perceptions. For, in the normal course of development, the child's understanding of the world becomes seemingly ever more sophisticated. The child is able to distinguish even finer gradations of 'goodness' and 'badness' so that by the later Oedipal stage an illusion of versimilitude is cast over the inherent (and irrational) distinction between the 'good' and 'bad' world and self, between control and loss of control, between acquiescence and denial.[3]

Both the self and the world are split into good and bad objects, and the bad self, the self associated with fear and anxiety over the loss of control, is projected onto bad objects. Fear precedes the construction of the bad object, the negative stereotype, but the stereotype – simplified, distorted and at a distance –perpetuates that fear. Most personalities draw on a range of stereotypes, not necessarily wholly good, not necessarily wholly bad, as a means of coping with the instabilities which arise in our perceptions of the world. They make the world seem secure and stable. While both good and bad stereotypes serve to maintain the boundaries of the self, to protect the self from transgressions when it appears to be threatened, most people have a large and sophisticated array of objects to draw on. As Gilman reminds us:

Our Manichean perception of the world as 'good' and 'bad' is triggered by the recurrence of the type of insecurity which induced our initial division of the world into 'good' and 'bad'. For the pathological individual, every confrontation sets up this echo . . . for the non-pathological individual, the stereotype is a momentary coping mechanism, one that can be used and then discarded once anxiety is overcome.[4]

It is evident that good and bad both resonate in stereotypical representations of others. As Zygmunt Bauman commented on taboos, which is what many stereotypes are, 'the human attitude is an intricate mixture of interest and fear, reverence and abhorrence, impulsion and repulsion'.[5] Thus, the stereotype may capture something that has been lost, an emotional lack, a desire, at the same time that it represents fear or anxiety. The good stereotype may represent an unattainable fantasy whereas the bad stereotype may be perceived as a real, malign presence from which people want to distance themselves. A common good stereotype of Gypsies, for example, locates them in the past or in a

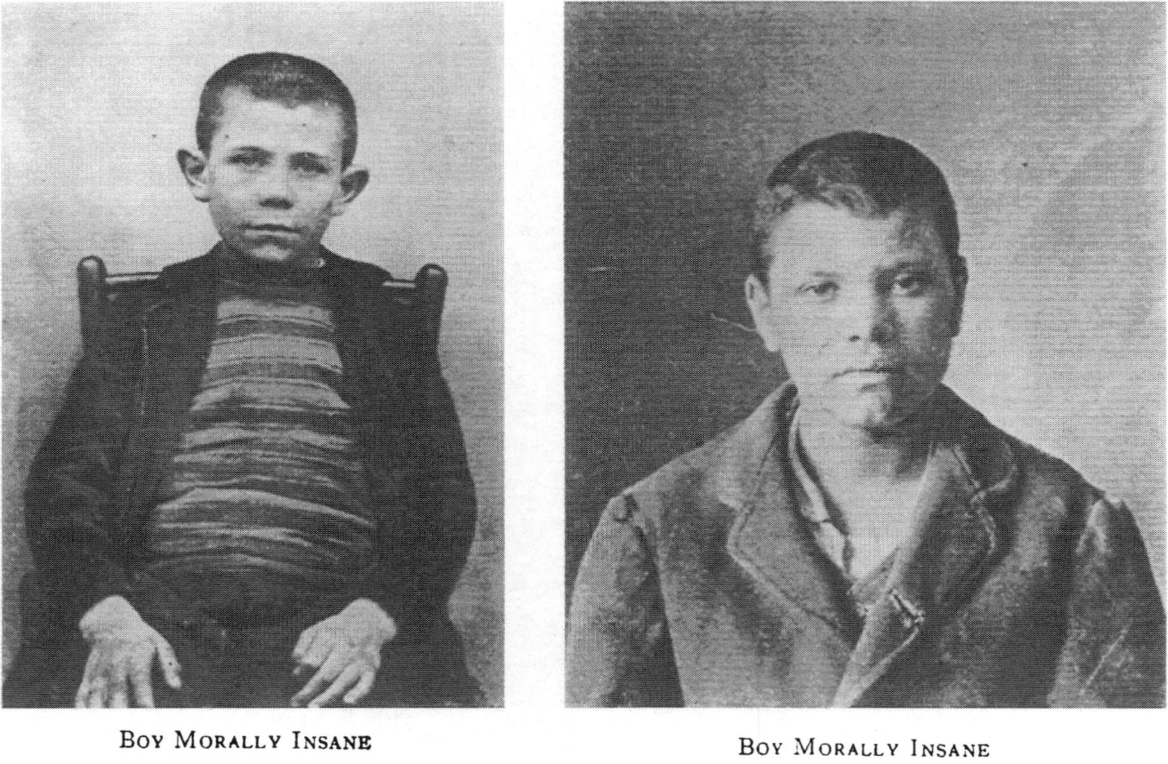

Boy Morally Insane

Boy Morally Insane

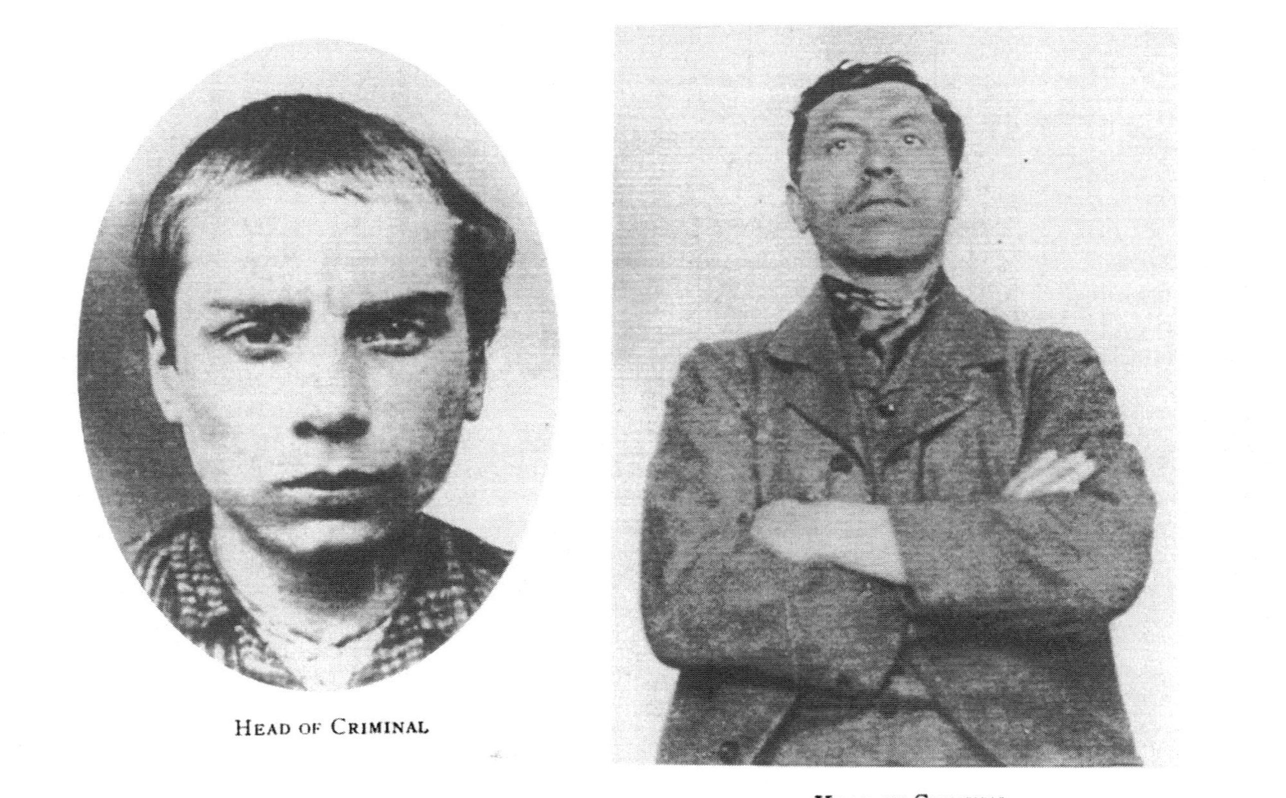

HEAD OF CRIMINAL

HEAD OF CRIMINAL

Plate 2.1 'Criminal types' from Cesare Lombroso's collection. Lombroso's use of photographic portraits in his work on criminality and madness demonstrates the historical importance of physical categorization in the cultural construction of normality and deviance (source: Ferrero 1911)

distant country where they are seen through a romantic mist. This is convenient because the good stereotype does not then contradict the bad stereotype. The Gypsy as a 'good object', an association of Gypsies with desire, is conveyed nicely in Hermann Hesse's poem *Glorious World*:

Sultry wind in the tree at night, dark Gypsy woman
World full of foolish yearning and the poet's breath.

Compare this with a characterization of Gypsies by Gina Ferrero, the daughter of the racist anthropologist Cesare Lombroso, in a commentary on her father's writing:[6]

an entire race of criminals, with all the passions and vices common to delinquent types: idleness, ignorance, impetuous fury, vanity, love of orgies and ferocity.

Both a fear for the boundaries of the self and a desire to merge are intimated in these representations, but in fact both dehumanize and contribute to a deviant image because both are, by definition, distortions. As Homi Bhabha suggests,[7] the stereotype is a simplification because it is an arrested, fixated form of representation which denies the play of difference. 'Others' disturb the observer's world-view, but the stereotype removes them from the scene in the sense that they are distinct from the world of everyday experience. Because there is little or no interaction with 'others', the stereotyped image, whether 'good' or 'bad', is not challenged.

Obviously, it is negative stereotypes which are of greatest consequence in understanding instances of social and spatial exclusion. Here, Julia Kristeva's conception of abjection, that unattainable desire to expel those things which threaten the boundary, and the abject, that list of threatening things and threatening others, seems to me to be fundamental. The earliest experience of abjection in the child is a reaction to excrement as the infant is socialized into adult categorizations of the pure and the defiled, and this then becomes a metaphor for other sources of defilement which are embodied in stereotypes. The sources of bodily defilement are projected onto others, whose world is *epidermalized*. As Iris Young argues:

When the dominant culture defines some groups as different, as the Other, the members of these groups are imprisoned in their bodies. Dominant discourse defines them in terms of bodily characteristics and constructs those bodies as ugly, dirty, defiled, impure, contaminated or sick.[8]

Or, as Stallybrass and White put it succinctly: 'Differentiation depends upon disgust.'[9] Verbal and visual images which have their source in the idea of defilement shade into those which represent the body as less than perfect. Thus, the photographs in Plate 2.1, which come from Cesare Lombroso's catalogue of the other,[10] point to a connection between visual images of physical imperfection, according to his scale of being which differentiates the normal and the deviant, and mental illness or disability, conditions which threaten the boundaries of the self. An obsession with scaling and measurement of physical characteristics in order to determine moral boundaries and marginalize the other was particularly characteristic of nineteenth-century and early twentieth-century science, but the association of appearances and moral characteristics is an enduring one.

I suggested in the last chapter that the social self could also be seen as a place-related self, and this applies also to stereotypes of the other which assume negative or positive qualities according to whether the stereotyped individual or group is 'in place' or 'out of place'. The cases that I discuss later in the book demonstrate how this condition of being discrepant or, conversely, of belonging, is identified. The issue concerns the extension of the 'generalized other' to things, material objects and places. Thus, a place stereotype might be a romantic representation of a landscape to which a social group are seen to belong or not, depending on the consistency or incongruity of the group and place stereotypes. For the moment, I will focus on some of the main signifiers of otherness in western cultures as a prelude to an integration of the social and spatial dimensions of the problem. The key areas that I examine are those of colour, disease and nature.

BLACK AND WHITE

The use of colour to signify positive/negative, life/death, superior/inferior, safe/dangerous, and so on, is evident in all cultures. Here, I will refer only to the use of colour in white European cultures, and then only black and white, because a cross-cultural account of the use of a wide palette of colours would be long-winded and not particularly relevant. The justification for considering these two colours is, first, that European nations are implicated in most accounts of racism and colonialism and rules expressed in terms of black and white have been important in the process of regulating and dominating the colonized and, second,

Plates 2.2a and b Dirt and blackness as signifiers in a white society. Moral instruction by the Health and Cleanliness Council, London (probably 1920s)

CHILDREN KNOW WELL THEIR TEACHER'S ATTITUDE TOWARDS DIRT.

that they are readily associated with defilement and purity.

Black is used routinely to describe dirt which, in turn, is associated with shame and disease. In other words, it has both practical and moral associations, which make it a potent marker of social difference. In the common usage of white Europeans, it is a negative signifier of class, race, ethnicity. The way in which black has been used to indicate class difference is suggested by the illustrations in Plate 2.2a and 2.2b, which come from a teachers' guide to health and cleanliness (published by the Health and Cleanliness Council in London in the 1920s and written by two Ph.D.s, one of whom was also the author of *Psychoanalysis in the Classroom*!). Black is also the colour of death, a source of defilement, a state which threatens life, and of the corpse, which signifies decay and contagion: 'If dung signifies the other side of the border, the place where I am not and permits me to be, the corpse, the most sickening of all wastes, is a border that has encroached upon everything.'[11] It is notable that organizations that have been devoted in a big way to death, the extreme rejection of difference, like the Nazis and other fascists, have adopted black for their collective identity.

Black, then, has been used in white societies to signal fear. A clear example of the use of black and associated images to convey a threatening otherness appears in Emily Brontë's *Wuthering Heights*, in which, as Rosemary Jackson notes,

> The family excludes everything foreign to itself as being unnatural. It guarantees ontological stability through limitation and closure. By the end of *Wuthering Heights*, the threat represented by Cathy and Heathcliff has been exorcised by confining it to their own vampiric relationship: they are merely restless spirits drifting around the abandoned closure of the Heights.[12]

The fear instilled in the family by Heathcliff depended on his portrayal from the beginning as other, as an outsider. Thus, as he came into the family (chapter 4):

> We crowded round, and over Miss Cathy's head I had a peep at the dirty, ragged, black haired child . . . I was frightened, and Mrs Earnshaw was ready to fling it out of doors. She did fly up, asking how he could fashion to bring that gipsy brat into the house, when they had their own bairns to feed and fend for?

Black-haired, dirty, Gypsy combine to suggest a threatening difference, drawing on an ethnic stereotype well established in British culture.

Black and white as racial signifiers have deep significance. In white, former colonial societies, as Dyer observes, there are inevitable associations of white with light, and therefore safety, and black with dark and therefore danger, and . . . this explains racism (whereas one might well argue about the safety of the cover of darkness and the danger of exposure to light).[13]

In a colonial context, black and white represent a whole set of social characteristics and power relations. Dyer's conclusion about three films portraying the white presence in colonial Africa – *Jezebel* (USA, 1938), *Simba* (Britain, 1955) and *Night of the Living Dead* (USA, 1969) – is that 'they all associate whiteness with order, rationality, rigidity, qualities brought out by contrast with black disorder, irrationality and looseness'.

This use of white and black is clearly intended to make white social behaviour virtuous and to legitimate white rule. However, white people should consider the question also from a black perspective. bell hooks argues that whites have a deep emotional investment in the myth of 'sameness' even as their actions reflect the primacy of whiteness as 'the sign informing who they are and how they think'.[14] In other words, whites do not think about themselves as white but only about others as not-white and other, which was the point of Dyer's examination of whiteness in films about colonialism. It is useful then to compare the dominant white view of blackness with black experience, like bell hooks's observations on the meaning of whiteness in her own childhood:

Returning to memories of growing up in the social circumstances created by racial apartheid, to all black spaces on the edge of town, I reinhabit a location where black folks associated whiteness with the terrible, the terrifying, the terrorizing. White people were regarded as terrorists, especially those who dared to enter the segregated space of blackness.[15]

An-other voice? White has been normalized in Europe, North America and Australasia and, in order to recognize that what seems normal is also a symbol of domination, it is important to listen to and to appreciate black perspectives rather than, as bell hooks suggests white people do, '[travel] around the world in pursuit of the other and otherness' – a sobering comment on academic enthusiasm for difference.

Returning to the wider associations of blackness in white societies, the association between black and dirt, between dirt and disease, emphasizes the threatening quality of blackness. Removing blackness, injecting light, removes fear, but this fear may be a fear of others as much as a fear of darkness. Thus, Corbin maintained that Haussmann's plans for Paris in the mid-nineteenth century were designed to make the city *less dark*: 'His town-planning was partly aimed at eliminating the darkness at the centre of the city, where darkness stood for the foul-smelling environment of the poor, the smell of the poor' – and the poor themselves.[16] However, despite the common use of black to signify obscurity, shade, shadows, fear, misfortune, death and evil, it has not always been so. Thus, in mediaeval Europe, black knights were courageous; numerous black madonnas, as in Tarragona cathedral, Spain, and Czestochowa, Poland, were objects of reverence. The positive associations of blackness were lost with colonialism, however, and with industrialization and the development of the class system under capitalism black assumed wider significance through its association with dirt, disorder and the threat to the bourgeoisie posed by the working class.

In the same system of values, whiteness is a symbol of purity, virtue and goodness and a colour which is easily polluted. Since whiteness is often not quite white, it is something to be achieved – an ideal state of pure, untainted whiteness. Thus, white may be connected with a heightened consciousness of the boundary between white and not-white, with an urge to clean, to expel dirt and resist pollution, whether whiteness is attributed to people or to material objects. As Sassoon remarks, white 'has a highly accentuated hygienic symbolism', although, in consumer culture, there has been to some extent an 'emancipation from white [which] has come about after several decades of emblematic monochromatism [*sic*]'.[17] As a marker of the boundary between purified interior spaces – the home, the nation, and so on – and exterior threats posed by dirt, disorderly minorities or immigrants, white is still a potent symbol.

DISEASE

Disease, often in combination with other signifiers of defilement, has a role in defining the self and in the construction of stereotypes. Its is a mark of imperfection and carries the threat of contagion. Disease in general threatens the boundaries of personal, local and national space, it engenders a fear of dissolution, and, thus, we project the fear of our own collapse, or of social disintegration, 'onto the world in order to localize it and . . . to domesticate it'.[18] Thus, the 'diseased other' has an important role in defining normality and stability.

Disease metaphors were characteristic of nineteenth-century scientific discourses which attempted to harness scientific knowledge in support of racist myths. Disease represented a threat of contagion, often a coded term for racial mixing, but to be diseased or disabled was a mark of imperfection. Western science was notable for its concern with classification and, particularly in the human sciences, with charting the contours of normality. Visual representations were particularly important for measuring and classification and the possibilities were increased considerably by photography. The case of Cesare Lombroso, engaged in cataloguing physical characteristics which were supposed to signify mental disability, mental illness, immorality and criminality, one slipping easily into the other, was not exceptional. This kind of scientism lent itself to racist theory and practice in Nazi Germany where Jews, defined in terms of physical difference or imperfection, became dangerous, like the mentally disabled, because of the threat they posed to the purity and stability of the Aryan race. The fascists engaged in 'object substitution', the threat personified interchangeably by Jews, women, homosexuals, communists, and so on. The disease metaphor, because it is universally threatening, was an appropriate one to elide with these demonized others.

Disease is a more potent danger if it is contagious. The fear of infection leads to the erection of the barricades to resist the spread of diseased, polluted others. The idea of a disease spreading from a 'deviant' or racialized minority to threaten the 'normal' majority with infection has particular power. This is apparent in current anxieties about AIDS, which reinforce homophobic or racist attitudes – AIDS as the gay disease, AIDS as the black African disease. A variant of this is evident in the Russian parliament's decision in 1994 to require all foreigners entering the country to have an AIDS test, in an attempt to 'seal [the country's] borders to "dirty" foreign bodies' (the *Guardian*, 12 November 1994). Here, the threat is embodied in the decadent West.

In the past, fears about class were expressed in similar terms. Contagious diseases like cholera or venereal disease were 'working-class diseases' which threatened the bourgeoisie and threatened to invade bourgeois space. A fear of infection was a fear of sexual touch; venereal disease, like AIDS, violated personal boundaries. It should be noted, however, that the presumed source of infection, the working-class prostitute, was also an object of desire. Disease, the working class and dirt were closely associated in nineteenth-century moralizing discourse. Today, disease, homosexuals and Black Africans have been similarly bracketed together. It is important to have somewhere (else) to locate these threats. It is a necessary part of distanciation. Thus, Africa and, in the recent past, San Francisco, have served as convenient depositories for threatening disease and diseased others, although hostile communities may simply be concerned with distancing themselves from the threat of infection by wishing 'diseased others' elsewhere.

NATURE

Nature has a long historical association with the other. Imperial science and theology both established hierarchies of being which put white civilization at the top and black people below, with groups like Australian Aborigines at the bottom because they were assumed to be a part of nature. Fitting species and human groups into taxonomic schemes was a major concern of nineteenth-century European science. Thus, Francis Galton, one of the founders of statistical analysis, put 'the average standard of the Negro race . . . two grades below our own; that of the Australian native . . . at least one grade below the African'.[19] Science confirmed the global dominance of white societies, a dominance which the church in colonial powers like Britain also asserted with its argument that peoples closest to nature, in a primitive state, needed saving. Salvation often involved not only accepting Christianity but also adopting European styles of dress and the discipline of a Christian education in the mission school. This was the fate of indigenous peoples in Australia and Canada, for example. The civilizing mission distanced them from nature. Brody noted that until quite recently the white stereotype of the Inuit in the Canadian Arctic placed them in the wild, as a part of nature, because of their capacity to withstand extreme cold and survive in a harsh environment by hunting and gathering.[20] While this image is a romantic one, projecting onto the Inuit what the white, urban Canadian has lost, it also serves to dehumanize. If they are a part of nature, they are less than human.

In a discussion of patriarchy, Capra made a similar observation in relation to the oppression of women. He argued that:

> Under patriarchy, the benign image of nature changed into one of passivity, whereas the view of nature as wild and dangerous gave rise to the idea that she was to be dominated by man. At the same time, women were portrayed as passive and subservient to men. With the rise of Newtonian science, finally, nature became a mechanical system that could be manipulated and exploited, together with the manipulation and exploitation of women. The ancient association of women and nature thus interlinks women's history and the history of the environment.[21]

Gillian Rose discusses the feminization of nature, the merging of women and nature, and the converse of this, the appropriation of culture by men, at some length.[22] However, observations on the representation of indigenous peoples and colonized others suggest to me that the nature association is not a peculiar characteristic of patriarchy, as Rose seems to argue, but is a more general feature of the scaling of beings by dominant groups which is closely associated with the history of colonialism, the rise of science and the growth of capitalism. Relegating dominated groups to nature – women, Australian Aborigines, Gypsies, African slaves – excludes them from civilized society. However, the inclusion of women in this list depends on their situation. In a colonial context, for example, civilized society would include those women who belonged to the colonial power although, at the same time, they would be subject to exclusions which result from the exercise of patriarchal power. Privileged membership of a dominant group or exclusion from it can only be explained in the particular contexts of race, ethnicity, gender or sexuality, as Judith Okely perceptively observed in her early essay on difference in Gypsy communities.[23]

The relegation of some groups to nature, where they are 'naturally' wild, savage, uncivilized, is also expressed in the representation of people as animals, either as animals generically distinct from humans or as particular species which are associated with residues or the borders of human existence – animals as the abject. Here, we find some of the most imaginative but also some of the most destructive stereotypical representations of others. Constance Perin notes that

> In the Great Chain of Being, humans are one mammal among the many. . . species of mammal. We are different from others yet not wholly so and, from that worrisome ambiguity, less-than-perfect human beings are perceived as resembling animals the more and placed below 'perfect' people in the social order.[24]

To dehumanize through claiming animal attributes for others is one way of legitimating exploitation and exclusion from civilized society, so it is unsurprising that it is primarily peripheral minorities, indigenous and colonized peoples, who have been described in these terms.

Mayall makes this observation about Gypsies, a minority who were subject to very harsh laws, including transportation and execution, in several European states in the eighteenth and nineteenth centuries:

> Perhaps the most overtly antagonistic and antipathetic of all the images of a race of Romanies was the likening of people to animals: 'The Gypsies are nearer to the animals than any race

known to us in Europe.' This statement appeared in an article entitled 'In Praise of the Gypsies'! The intention, then, was to place the Gypsies on the lowest possible level of human existence. They were said to eat more like beasts than men [*sic*].[25]

Rats, pigs and cockroaches[26] have had a particular place in the racist bestiary because all are associated with residues – food waste, human waste – and, in the case of rats, there is an association with spaces which border civilized society, particularly subterranean spaces like sewers, which also channel residues and from which they occasionally emerge to transgress the boundaries of society. The potency of the rat as an abject symbol is heightened through its role as a carrier of disease, its occasional tendency to violate boundaries by entering people's homes, and its prolific breeding. Thus, the rat has been readily adopted in racist propaganda such as anti-Semitic films produced by the Nazis which portrayed Jews as rats. Similarly, in England in the nineteenth century the rat was one stereotype of the Irish minority. As Stallybrass and White observed,

Once the metaphoric relations were established, they could be reversed. If the Irish were like animals, animals were like the Irish. One of the sewer workers . . . described the sewers (which Irish labourers had helped to build) as full of rats 'fighting and squeaking. . . like a parcel (!) of drunken Irishmen'.[27]

The Irish were seen as uncivil, belonging outside English civil society because of their association with dirt, like rats, both through their work on canals, sewers and railways and through their rural Irish background, 'where the imagined connection between peasants and dung was very close'. This dehumanization was, of course, a necessary part of the colonial relationship between Britain and Ireland. Otherwise, it is a very clear example of the way in which dirt, disease and nature are combined in an animal metaphor. Transgressive mammals and insects have played an important part in the construction of negative stereotypes.

CONCLUSION

This discussion has one important practical implication. In local conflicts, where a community represents itself as normal, a part of the mainstream, and feels threatened by the presence of others who are perceived to be different and 'other', fears and anxieties are expressed in stereotypes. However, engaging with the other, what bell hooks calls repositioning,[28] might lead to understanding, a rejection of a stereotype and a lesser concern with threats to the boundaries of community. Any optimism about such a move should be tempered with the thought that limited engagement, a superficial encounter, might result in the presumption of knowledge which could be more damaging than ignorance if this knowledge were in the province of state bureaucracies or academia. Thus, I would see repositioning as problematic. The acquisition of authentic knowledge also raises important methodological questions which I will consider at the end of the book.

More generally, I have been concerned in this chapter to demonstrate the link between object relations theory and images which play an important part in the construction of stereotypes. The good and bad objects of Kleinian theory are not catalogued in any detail but it should be clear that both the self and stereotypes are products of culture and society, so it is important to identify some of the images that stand for the other and to contextualize observations on selfhood and constructions of otherness. Verbal and visual images, as they are rooted in culture, are the things to which people, as individuals and as social beings, relate. I have begun to sketch in some of the stereotypes which have currency in western cultures, but in subsequent chapters these pictures will become more detailed and elaborate.

NOTES

1. Some recent post-modern writing, for example, Iain Chambers's *Migrancy, Culture and Identity*, Routledge, London, 1994, celebrates difference with some enthusiasm. The theme of Chambers' book is that there are fusions, hybrids and new forms of difference that follow from increasing global movement and interconnectedness. I think that it is important not to be carried away by this. Problems defined by the firm contours of territorially based conflict, associated with race, ethnicity, sexuality and disability, are persistent features of socio-spatial relations. Many people live in one place for a long time and some have difficulty getting along with those who are different from themselves. Unfortunately, the African musicians whom Chambers admires and who have certainly enriched British culture are still subject to racism outside the sympathetic environment of the club or the music festival.

2. Sander Gilman, *Difference and Pathology: Stereotypes of sexuality, race and madness*, Cornell University Press, Ithaca, N.Y., 1985. This book, with its emphasis on visual representation, develops object relations theory to incorporate the world as it is perceived.

3. ibid., p. 17.

4. ibid., p. 18.

5. Zygmunt Bauman, 'Semiotics and the function of culture', in Julia Kristeva *et.al.* (eds), *Essays in Semiotics*, Mouton, The Hague, 1971, pp. 279–295.

6. Gina Ferrero, *Criminal Man*, The Knickerbocker Press, New York, 1911, p. 140.

7. Homi Bhabha, *The Location of Culture*, Routledge, London, 1994. Bhabha presents a deep analysis of the 'colonial other', but his arguments have much wider relevance.

8. Iris Young, *Justice and the Politics of Difference*, Princeton University Press, Princeton, N.J., 1990, p. 126. Young bases her argument on Julia Kristeva's *Powers of Horror.*

9. Peter Stallybrass and Allon White, *The Politics and Poetics of Transgression*, Methuen, London, 1986, p. 191.

10. The subjects of Lombroso's moralizing discourse were primarily people with learning disabilities. The importance of photography as an aid to classifying mentally ill and disabled others is discussed in some detail by Sander Gilman, *Disease and Representation*, Cornell University Press, Ithaca, N.Y., 1988, pp. 39–43. Lombroso's photographs are reproduced in Ferrero, op. cit.

11. Julia Kristeva, *Powers of Horror*, Columbia University Press, New York, 1982, p. 3.

12. Rosemary Jackson, *Fantasy: The literature of subversion*, Methuen, London, 1981, p. 129.

13. Richard Dyer, *The Matter of Images: Essays on representation*, Routledge, London, 1993, pp. 142–145.

14. bell hooks, *Black Looks: Race and representation*, Turnaround, London, 1992.

15. ibid., p. 170. Later in this chapter (p. 174), she remarks:

> Reminded of another time when I was strip searched by French officials, who were stopping black people to make sure we were not illegal immigrants and/or terrorists, I think that one fantasy of whiteness is that the threatening Other is always a terrorist. This projection enables many white people to imagine there is no representation of whiteness as terror, as terrorizing. Yet it is this representation of whiteness in the black imagination, first learned in the confines of [a] poor black community, that is sustained by my travels to many different locations.

16. Alain Corbin, *The Fragrant and the Foul: Odor and the French social imagination*, Harvard University Press, Cambridge, Mass., 1986, pp. 134–135.

17. Joseph Sassoon, 'Colors, artefacts, and ideologies', in P. Gagliardi (ed.), *Symbols and Artefacts: Views of the corporate landscape*, de Gruyter, Berlin, 1990, p. 172.

18. Gilman, 1988, op. cit., p. 1.

19. Cited by C. P. Blacker, *Eugenics: Galton and after*, Duckworth, London, 1952, p. 325.

20. Hugh Brody, *The People's Land: Eskimos and whites in the eastern arctic*, Penguin, Harmondsworth, 1975. This romantic construction of the Inuit corresponded to the 'real Eskimo', as opposed to the welfare-dependent alcoholic of the bad stereotype.

21. Frijtof Capra, *The Turning Point: Science, society and the rising culture*, Bantam Books, London, 1987, p. 24.

22. Gillian Rose, *Feminism and Geography*, Polity Press, Cambridge, 1993, chapter 4.

23. Judith Okely, 'Gypsy women: models in conflict', in Shirley Ardener (ed.), *Perceiving Women*, Dent, London, 1975, pp. 55–86. Okely's research demonstrates the value of participant observation in discerning roles and power relations in a community which had formerly been portrayed as an undifferentiated other. Her writing anticipated Michelle Barrett's plea for the

deconstruction of categories like 'women' – Michelle Barrett, 'The concept of difference', *Feminist Review*, 26, 1987, 29–41.

24. Constance Perin, *Belonging in America,* University of Wisconsin Press, Madison, 1988, pp. 144–145.

25. David Mayall, *Gypsy-Travellers in Nineteenth Century Society*, Cambridge University Press, Cambridge, 1988, chapter 4.

26. Cockroach symbolism is discussed by Hoggett in a study of racism in the East End of London, where the Muslim Bangladeshi population constitutes the other for the working-class whites. The cockroach 'signifies a complex knot of resentment, fear and hatred', directed at the Bangladeshis, although, of course, cockroach infestations are caused by structural defects in their homes. P. Hoggett, 'A place for experience: a psychoanalytic perspective on boundary, identity and culture', *Environment and Planning D: Society and Space*, 10, 1992, pp. 345–356.

27. Stallybrass and White, op. cit., p. 133.

28. hooks, op. cit., p. 177.

3

BORDER CROSSINGS

The sense of border between self and other is echoed in both social and spatial boundaries. The boundary question, a traditional but very much under-theorized concern in human geography, is one that I will explore in this chapter from the point of view of groups and individuals who erect boundaries but also of those who suffer or whose lives are constrained as a result of their existence. Crossing boundaries, from a familiar space to an alien one which is under the control of somebody else, can provide anxious moments; in some circumstances it could be fatal, or it might be an exhilarating experience – the thrill of transgression. Not being able to cross boundaries is the common fate of many would-be migrants, refugees, or children in the home or at school. Boundaries in other circumstances provide security and comfort. I will start by examining some general characteristics of boundary zones and then describe some of the diverse ways in which boundaries are constructed, demolished and energized.

In a rather formalistic treatment of the boundary problem, Edmund Leach played with the idea of separating and combining categories, focusing particularly on the intersections of sets.[1] Like Leach, I will use Venn diagrams to introduce some of the boundary issues which are characteristic of social and spatial relations, issues which are central to questions of exclusion. The need to make sense of the world by categorizing things on the basis of crisp sets – A, not-A, and so on – is evident in most cultures, although I do not think it is a universal need, as Leach suggests. However, it is a good place to start because problems associated with this mode of categorization are readily identifiable.

Problems arise when the separation of things into unlike categories is unattainable. The mixing of categories, indicated in Figure 3.1 by the

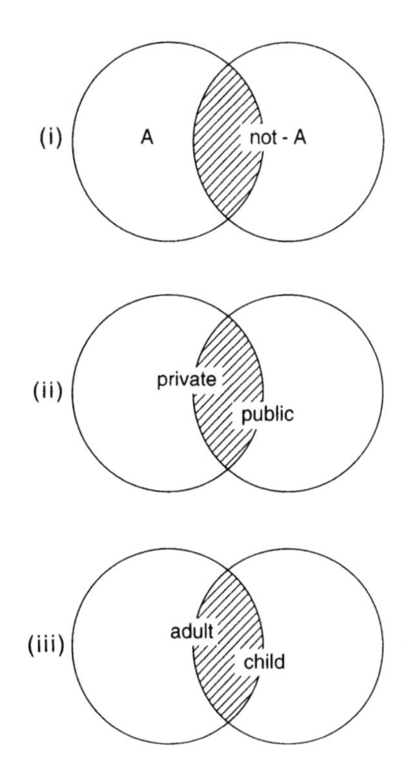

Figure 3.1 Zones of ambiguity in spatial and social categorizations. Danger, or at least uncertainty, lies in marginal states

intersections of sets, creates liminal zones or spaces of ambiguity and discontinuity. As Leach recognized, 'There is always some uncertainty about where the edge of Category A turns into the edge of Category not-A.'[2] For the individual or group socialized into believing that the separation of categories is necessary or desirable, the liminal zone is a source of anxiety. It is a zone of abjection, one which should be eliminated in order to reduce anxiety, but this is not always possible. Individuals lack the power to organize their world into crisp sets and so eliminate spaces of ambiguity.

To move from Leach's abstraction (Figure 3.1(i)) to the concrete, we might consider the case of the home (Figure 3.1(ii)). To the occupier, the home may represent a space clearly separated from the outside. Inside the home, the owner or tenant may feel that space is ordered according to his or her values.[3] However, problems can be created by entrances, breaches in the boundaries of the home. The entrance, the hallway or passage provides a link between the private and the public, but it constitutes an ambiguous zone where the private/

Plate 3.1 Teenagers 'out of place' on a children's playground (photo: Jo Goodey)

public boundary is unclear and in need of definition and regulation in order to remove the anxiety of the occupier.[4] If you admit strangers to the house, are they confined to an entrance area or allowed to enter a living space? How do you cope with the Jehovah's Witnesses or with the person selling double-glazing? The response will depend on where the householder locates the boundary, but this may be variable, depending on how the outsider is perceived in relation to the occupier's conception of privacy.

A second example, child/adult (Figure 3.1(iii)) illustrates a similarly contested boundary. The limits of the category 'child' vary between cultures and have changed considerably through history within western, capitalist societies. The boundary separating child and adult is a decidedly fuzzy one. Adolescence is an ambiguous zone within which the child/adult boundary can be variously located according to who is doing the categorizing.[5] Thus, adolescents are denied access to the adult world, but they attempt to distance themselves from the world of the child. At the same time, they retain some links with childhood. Adolescents may be threatening to adults because they transgress the adult/child boundary and appear discrepant in 'adult' spaces. While they may be chased off the equipment in the children's playground (Plate 3.1), they may also be thrown out of a public house for under-age drinking. These problems encountered by teenagers demonstrate that the act of drawing the line in the construction of discrete categories interrupts what is naturally continuous. It is by definition an arbitrary act and thus may be seen as unjust by those who suffer the consequences of the division.

BOUNDARY MAINTENANCE AND SOCIAL ORGANIZATION

In using these two examples, I am suggesting that liminality presents as many problems for highly developed capitalist societies as for the relatively simple agrarian and hunter-gatherer collectivities which have been the primary focus of anthropological research and where much of the theory of boundary dynamics has been developed. Dichotomies like traditional/modern or simple/complex do not seem to have much relevance to the questions of boundary drawing, inclusions and exclusions. Perhaps a meaningful distinction could be made between what Davis and Anderson term *high-density* and *low-density* social networks.[6] Albeit crude, this dichotomy does suggest varying attitudes to difference which might be attributed to the density of social interaction. Davis and Anderson suggest that in high-density networks 'most links are strong and one is likely to know and have direct ties to most people affected by the misbehavior [*sic*] of a member of one's network'. Conversely, they argue that in networks of low density, difference is less visible because there is less shared knowledge of individuals within the community. In pre-industrial societies, people are enmeshed, involved in each other's lives through extended family, kin connections or clan membership coupled, in many cases, with simple physical propinquity. Gypsies and other semi-nomadic minorities demonstrate the characteristics of shared knowledge of members of the community and of physical nearness very clearly. An outsider in such a community is very exposed. However, similar forms of social organization are found within developed societies, both within traditional working-class and suburban neighbourhoods.

A division based on the density of social networks is fairly close to Durkheim's schema in which he distinguishes between societies exhibiting *mechanical* and *organic* solidarity.[7] Where social identity is based on mechanical solidarity, Durkheim argues that

> the society is dominated by the existence of a strongly formed set of sentiments and beliefs shared by all members of the community [so] it follows that there is little scope for differentiation between individuals: each individual is a microcosm of the whole.

With shared beliefs, we can talk of a *conscience collective* which 'completely envelopes individual consciousness'.[8] By contrast, individualism is consistent with social solidarity in developed industrial societies because solidarity is organic, that is, deriving from contractual relationships which develop with an increasing division of labour. Durkheim assumed that these contracts were governed by norms which comprised the glue holding society together, but norms did not preclude individual difference. This dichotomy does considerable violence to reality and it was probably a fairly crude representation of varying forms of social organization when Durkheim was writing in the late nineteenth century. Like Davis and Anderson's view of social networks, it does provide some ideas about the way people might collectively react to difference, but we need knowledge of the social, political and geographical contexts of community responses to 'others' in order to say anything useful about conflicts based on difference. With the globalization of culture and the almost total penetration of capitalist forms of consumption, it certainly does not make sense to characterize societies in Durkheim's terms. Yet, there is something approaching a conscience collective in some middle-class North American suburbs[9] and on some local authority housing estates in Britain, manifest in reactions to the mentally disabled, Gypsies and Bangladeshis, for example. Some useful ideas on this kind of hostility to others comes from studies of small groups by social anthropologists. Here, the work of Mary Douglas and her critics is particularly illuminating.

POLLUTION, DISCREPANCY AND SMALL GROUP BOUNDARIES

At the social level, as at the individual level, an awareness of group boundaries can be expressed in the opposition between purity and defilement. In Mary Douglas's *Purity and Danger* and subsequent writing,[10] she developed this thesis, gathering support for her argument largely from fieldwork with tribal societies and ancient texts, particularly the Old Testament of the Bible as a record of Judaic ritual. Her key argument is that 'Uncleanness or dirt is that which must not be included if a pattern is to be maintained . . . in the primitive culture, the rule of patterning works with greater force and more total comprehensiveness [*sic*] . . . [than in a modern industrial society]'.[11] By patterning, Douglas means the imposition of a symbolic order 'whose keystone, boundaries, margins and internal lines are held in relation by rituals of separation'.[12] Separation is a part of the process of purification – it is the means by which defilement or pollution is avoided – but to separate presumes a categorization of things as pure or defiled. This can be illustrated by her analysis of a passage from Leviticus:

> The last kind of unclean animal is that which creeps, crawls or swarms upon the earth . . . Whether we call it teeming, trailing, creeping, crawling or swarming, it is an indeterminate form of movement. Since the main animal categories are defined by their typical movement, 'swarming', which is not a mode of propulsion proper to any particular element, cuts across the basic classification. Eels and worms inhabit water, though not as fish; reptiles go on dry land, though not as quadrupeds; some insects fly, though not as birds. There is no order in them.[13]

From such observations, Douglas proposes a rule for categorization in ancient Israel which forms the basis for a general rule, namely, that 'the underlying principles of cleanliness in animals is that they shall conform fully to their class. Those species are unclean which are imperfect members of their class or whose class itself confounds the general scheme of the world'.[14] Thus, it is those animals, people or things that are discrepant, that do not fit in a group's classification scheme, which are polluting. The evidence for this from records of the practices current in ancient Israel are quite convincing and Neusner has compiled a long list of polluting activities, conditions of the body, animals, and so on, which lends support to Douglas's thesis.[15] However, to generalize from one ancient culture, where there were strong rules of exclusion laid down by the rabbis who also wrote the texts that constitute the evidence, to all small groups and tribal societies is dangerous. Murray, in particular, questioned the empirical validity of her argument, suggesting that people were not really that concerned about defilement and happily mixed discrepant categories in their daily struggle for survival.[16] He asserted that: 'If there is any psychological reality to the "horror" purportedly inspired by such classification difficulties, it is confined to anthropologists intent on eliciting complete and exhaustive contrast sets.' According to Murray, the people who were supposedly engaged in these boundary rituals would be driven to anxiety 'by proscribing and even attempting to annihilate what is not readily classifiable from the world'. This criticism was hardly fair because Mary Douglas had herself recognized the limitations of her original thesis, long before Murray's assault.[17]

I think we can conclude that Douglas's argument about purification and defilement needs to be qualified in regard to time and place, but I would also argue that it has wider application than she recognized. In *gemeinschaft*-like groups, that is, closed, tightly knit communities with something approaching a conscience collective, it may be that adherence to the rules is more likely in times of crisis, when the identity of the community is threatened. However, my observations in English Gypsy communities suggest that poverty and family size are also factors affecting observance of pollution taboos. In order to ensure that things are not polluted (*mochadi* or *marime*), numerous separations are required, including among utensils which are used for washing food, clothes, and the body. In poor families with several small children, however, it is often impractical to comply with all these rules. There may be a shortage of water or not enough washing bowls and the mother may be too tired to meet the ritual requirements all the time. In affluent families, particularly with grown-up children, pollution taboos are much more likely to be observed. There is some support here for Murray's argument, but he overstates his case.

Boundary consciousness is also a characteristic of the mainstream in modern, western society, or, at least, it is in some kinds of locales and at certain times. The North American suburb has been represented as a particular kind of *gemeinschaft* within the swathe of individual anonymous worlds that are supposed to constitute the modern metropolis. The suburb was first described as an exclusionary, purified social space by Richard Sennett, in *The Uses of Disorder*;[18] the anatomy of the North American variety has been examined in some detail by Constance Perin;[19] and Mike Davis describes the enclosed communities, socially purified and defended fortress-like, against the supposed threat posed by the poor, which are an increasingly prominent feature of the geography of Los Angeles.[20] Affluent suburbs in Britain are similarly coming to resemble these closed communities where the discrepant is clearly identified and expelled, like a suburb in Bristol which hired a private security firm to patrol its leafy avenues and eject what the security officers described as 'hostiles', including young men in baseball caps.

In these suburbs, there is a concern with order, conformity and social homogeneity,[21] which are secured by strengthening the external boundary, but, as Davis recognizes, 'the greater the search for conformity, the greater the search for deviance; for without deviance, there is no self-consciousness of conformity and *vice-versa*'.[22] This process is seen by the members of the community as a virtuous one – it brings into being a morally superior condition to one where there is mixing because mixing (of social groups and of diverse activities in space) carries the threat of contamination and a challenge to hegemonic values. Thus, spatial boundaries are in part moral boundaries. Spatial separations symbolize a moral order as much in these closed suburban communities as in Douglas's tribal societies.

BOUNDARY ENFORCEMENT

Generalizations like Sennett's 'purified suburb' have to be qualified. While there is plenty of evidence that purified suburbs exist, with damaging consequences for the welfare of the rest of the population in metropolitan areas, not all suburbs are like this. Apart from an increase in racially mixed suburbs in the United States,[23] it has also been argued that (British) suburbs can provide a refuge for eccentrics.[24] A concern with privacy, minding your own business, is also characteristic suburban behaviour. However, communities which much of the time appear to be indifferent to others do occasionally turn against outsiders, particularly when antagonism is fuelled by moral panics. Moral panics heighten boundary consciousness but they are, by definition, episodic. Fears die down and people subsequently rub along with each other. Often, but not invariably, panics concern contested spaces, liminal zones which hostile communities are intent on eliminating by appropriating such spaces for themselves and excluding the offending 'other'.

In Stanley Cohen's classic *Folk Devils and Moral Panics*, he decribes the phenomenon as

> a condition, episode, person or group of persons [which] emerges to become defined as a threat to societal values and interests: its nature is presented in a stylized and stereotypical fashion by the mass media . . . ways of coping are evolved or (more often) resorted to; the condition then disappears . . . or deteriorates and becomes more visible.[25]

One of the most remarkable features of moral panics is their recurrence in different guises with no obvious connection with economic crises or periods of social upheaval, as if societies frequently need to define their boundaries. As Cohen reflected on the Mods and Rockers episode in the 1960s,

> More moral panics will be generated and other, as yet nameless, folk devils will be created. This is not because such developments have an inexorable inner logic but because our society as presently structured will continue to generate problems for some of its members – like working-class adolescents – and then condemn whatever solutions these groups find.

The resonance of historical panics in modern crises is worth noting because it demonstrates the continuing need to define the contours of normality and to eliminate difference. This is as evident in what are claimed to be post-modern western societies, post-modern in the sense that they embrace difference,[26] as it was in Nazi Germany or England in the seventeenth century. I will describe one case from seventeenth-century England because of its remarkable similarity with modern panics, particularly in the way in which difference is represented by the popular media.

In the early modern period, before the industrial revolution in Europe and North America, Christian religion was an important source of conflict. The line between conformity and dissent, good and evil, light and darkness was of great concern for an established church attempting to isolate or eliminate religious minorities which appeared to threaten its theological hegemony. The Ranters were one such group who appeared threatening despite their very small numbers.[27] Their dissent from the established church was not unique but their views were publicized at a critical juncture in English history, just after the Revolution of 1649 and the establishment of the Commonwealth, when there was general political uncertainty. At this time, religion was politics and the collapse of religious authority was portrayed in sensationalist literature as 'a prelude to unbridled immorality and social chaos'.[28]

In this context, the Ranters represented a subversive and threatening group and their difference and deviance were amplified in a sustained press campaign which had little regard for the truth. Connections were suggested between religious dissent, atheism and immorality. Interestingly, as in witch crazes, Ranters were associated with inversions which threatened moral values, specifically, devil worship and promiscuity. While women generally had little scope for sexual relations outside marriage, Ranter women were portrayed as sexually unbounded: 'They were free to copulate with any man and did so enthusiastically and openly'. One broadsheet, 'Strange NEWES from the OLD-BAYLY', asserted in terms worthy of a modern tabloid that

Some [Ranters] have confessed that they have had often meetings, whereat both men and women presented themselves stark naked one to the other, in a most beastly manner. And after satisfying their carnall and beastly lust, some have a sport called whipping the whore; others call for musick and fall to revelling and dancing.[29]

The broadsheet had a major role in developing the Ranter stereotype and in fostering a hysterical response to the minority, accentuating their deviance and so legitimating oppression. Modern analogies are cases of ethnic conflict where the media have an important role in demonizing the enemy, as in the use of the radio by the Hutu to generate opposition to the Tutsi in Ruanda, although cases like this, or the ethnic conflicts in the former Yugoslavia, are more enduring, not panics in Cohen's terms.

The Ranter panic is comparable to modern instances in the sense that the group is represented as a threat to core values. Core values in the seventeenth century were to a great extent religious values, whereas in modern secular societies, core values are embodied in the family, the home and the nation, and thus they have implications for 'deviant' youth, other sexualities and racial minorities. Although the folk devil comes in different guises, panics do not introduce a succession of new characters bearing no resemblance to each other. Rather, they are manifestations of deep antagonisms within society, for example, between adults and teenagers, blacks and whites, heterosexuals and homosexuals. The alterity personified in the folk devil is not any kind of difference but the kind of difference which has a long-standing association with oppression – racism, homophobia, and so on. The moral panic will be accompanied by demands for more control of the threatening minority, for the state to provide stronger defences for, say, white, heterosexual values. This may include a call for the stronger bounding of space to counter the perceived threat, as in attempts by the British government to exclude New Age Travellers from the countryside and secure this terrain for the middle classes.[30] An account of two recent sources of moral panics in North America and Britain – AIDS and mugging – demonstrates most of these points.

Media responses to AIDS went through a period in the 1980s when they were typically moralizing and exaggerated.[31] There was a moral panic, beginning in the early 1980s and lasting until the late 1980s, after which there was more restrained and informed reporting, at least, a more muted homophobia and racism when it was realized that AIDS was not a 'gay plague' or an 'African disease' but also affected the white, heterosexual population. The wilder threats of contagion could not be sustained as knowledge about the syndrome and advice on safer sexual behaviour were more widely disseminated. However, during the 1980s, there was a 'spiralling escalation of the perceived threat, leading to a taking up of absolutist positions and the manning of moral barricades'.[32] The moral barricades were manned on behalf of 'the family'. As Simon Watney comments, the press made its primary appeal to the family as the central site of consumption, thus of fundamental importance to the economy, and as the site of sexuality and child rearing – 'The family is positioned in newspaper discourse as [a] central term of professional journalistic know-how, establishing a fixed agenda of values, interests and concerns which are heavily moralized.' The family was threatened by alternative models presented by gays and by AIDS as a 'gay disease'.

One panic response in western societies was to advocate quarantine, the physical isolation of homosexuals with AIDS, a response which betrays ignorance of the epidemiology of the syndrome but is consistent with the idea of the threat of contagion, a metaphoric threat to the health of the family. A spokesperson for the right-wing Conservative Family Campaign in Britain advocated the removal of 'AIDS victims' to quarantine centres. A similar proposal in Australia indicated to one critic that the panic was not only 'a product of homophobia but was . . . tied to the belief that [Australians] can insulate themselves from the rest of the world through rigid immigration and quarantine laws'.[33] Family space, threatened by other sexualities, has its counterpart in national space, threatened by alien values. Homophobia will not go away while homosexuality is constructed as an 'other' which threatens the boundaries of the social self. AIDS, however, was the catalyst for a panic which temporarily reinforced boundaries. It brought homophobia into sharp relief.

Sexuality is a source of difference from which moral panics can emerge because it is fundamental to people's world-views and their relationships to others. Similarly, race, as it is culturally constructed, can be a source of social cleavage which can be magnified in an episodic fashion. Thus, a mugging episode in Britain in the 1970s emerged from the problem of racism, which is deeply rooted in the centres of former empires. The particular target of moral outrage in this case was young Afro-Caribbean males.

Two things came together in the reporting of street crime in the 1970s. The first was the notion of the British inner city as a *black* inner city, characterized by lawlessness and vice, so that inner city became a coded term for black deviance. The idea of a black inner city bore little relation to demographic or geographical reality, but the myth is more important than the reality. Thus, in a typical example of place labelling, Weaver claimed that the local press 'portrayed north central Birmingham as a violent, crime-ridden area, beset by problems rooted in the nature of its coloured [*sic*] residents rather than in the district's disadvantaged position in British urban space'.[34] Second, there was a rise in recorded street crime but a perceived rise, particularly in 1973, amounting to 'a national mugging scare',[35] and this fixed the idea of black youth as an inherently criminal minority and inner cities as inherently criminal localities. Susan Smith's quotations from Birmingham newspapers, like 'Society at limit of leniency' and 'Angry suburb', indicate a panic which was generated through the stereotyping of minority group and locality. This required a silence about policing, unemployment, the population composition of the district and comparative crime statisitics for the city, which would have put a different complexion on the issue. The labels attached to the inner city during this panic strengthened the boundary separating the 'respectable white suburbs' and 'black inner areas' and decreased the likelihood of white people gaining knowledge of Afro-Caribbean and Asian communities through experience. A panic surrounding inner-city riots in the 1980s again confirmed the boundary, with material consequences for inner-city residents, such as the withdrawal of financial services. Different moral panics with slightly different scripts signalled the continuing presence of racism.

'Family', 'suburb' and 'society' all have the particular connotation of stability and order for the relatively affluent, and attachment to the system which depends for its continued success on the belief in core values is reinforced by the manufacture of folk devils, which are negative stereotypes of various 'others'. Moral panics articulate beliefs about belonging and not belonging, about the sanctity of territory and the fear of transgression. Since panics cannot be sustained for long, however, new ones have to be invented (but they always refer to an old script).

INVERSIONS AND REVERSALS

Moral panics bring boundaries into focus by accentuating the differences between the agitated guardians of mainstream values and excluded others. Occasionally, these social cleavages are marked by inversions – those who are usually on the outside occupy the centre and the dominant majority are cast in the role of spectators. Inversions can have a role in political protest in the sense that they expose power relations by reversing them and, in the process, raise consciousness of oppression. They energize boundaries by parodying established power relations.

In early modern Europe, inversions constituted a popular genre known as World Upside Down.[36] Broadsheets illustrating such a world were widely distributed among the illiterate, the themes being virtually unchanged for several hundred years. Illustrations showed, for example, the blind leading the sighted, sheep eating wolf, child punishing father, beggar giving alms to the rich. Their popularity with the oppressed could be accounted for by the fact that they fantasized about the existing order. Sometimes, reversals could serve as a symbol of actual revolt. In the Rebecca riots against the turnpike roads in Wales and the west of England, for example, men dressed as women and took Rebecca as a symbol of power against authority:

> And they blessed Rebecca and said unto her, 'Thou art our sister: be thou the mother of thousands of millions, and let thy seed possess the Gates of those who hate them.'[37]

This protest demonstrated the particular importance of 'women on top' as a symbolic reversal. It has a long history in Europe and Davis has suggested that

> males drew upon the sexual powers and energy of [unruly women] and their licence (which they had long ago assumed at carnivals and games) to promote fertility, to defend the community's interests and standards and tell the truth about unjust rule.[38]

The occasions when inversions assume a centre–periphery form, when the dominant society is relegated to the spatial margins and oppressed minorities command the centre, may represent a challenge to established power relations and, thus, be subject to the attentions of the state. There may be attempts to control or suppress such events because they harness the energies of groups which challenge mainstream values. As Stallybrass and White observe,

> There is no reason to suppose that capitalism should be . . . different from other societies in locating its most powerful symbolic repertoires at borders, margins and edges, rather than at the accepted centres of the social body.[39]

This is particularly the case with carnivalesque events which are licensed but have contested spatial and temporal bounds. For example, Caribbean carnivals in British cities have been grudgingly accepted by the state as legitimate celebrations of black culture in an avowedly pluralist society,[40] but, in the past, they have been heavily policed and contained. The appeal of the exotic for the white majority mixes uneasily with the images of black criminal stereotypes which have informed the responses of the control agencies. Similar conflicts occurred over the carnivalesque centre on festivals in rural England which attract New Age Travellers. With the Criminal Justice and Public Order Bill,[41] the British government is attempting to seriously limit or ban festivals which are seen as a threat to the cherished values of rural England. Inversions of this kind are thus important indicators of marginality. Responses to carnivalesque events demonstrate how the majority constructs the 'other'.

Other reversals may have less political currency although they can still be symbolically potent. One such case is the Gypsy pilgrimage to Saintes Maries de la Mer on the Camargue coast in the south of France. Gypsies have been relegated to the margins of French society for centuries, and being a Gypsy in the seventeenth and eighteenth centuries was under most regimes a crime warranting execution, mutilation, transportation or a life sentence in the galleys.[42] Since 1935, however, French and other European Gypsies have taken over the small town of Saintes Maries de la Mer on 24 and 25 May and again in October, for a ritual which inverts the practice of the established church in that the object of reverence is Sara, a black madonna. It was only in 1935 that French Gypsies gained ecclesiastical authority to venerate Sara, who has not been canonized by Rome. Although the pilgrimage has now been given a tourist gloss and the Gypsy veneration includes the other Saintes Maries, Salome and Jacobe, it is still a subversive event which expresses the collective but highly circumscribed power of European Gypsies and expresses the long history of racism to which they have been subject.

CONCLUSION

The propositions of object relations theory – the bounding of the self, the role of good and bad objects as stereotypical representations of others, as well as their representation as material things and places – can be projected onto the social plane. The construction of community and the bounding of social groups are a part of the same problem as the separation of self and other. Collective expressions of a fear of others, for example, call on images which constitute bad objects for the self and thus contribute to the definition of the self.

The symbolic construction of boundaries in small groups which have been studied by social anthropologists has its counterpart in the marking off of communities in developed western societies. Consciousness of purity and defilement and intolerance of difference secure some groups within the larger spaces of the modern metropolis. The outside is populated by a different kind of people who threaten disorder, so it is important to keep them at a distance. These fears, however, are fuelled by the exaggerated accounts of some sections of the media and the state who represent the claims of others for space, or simply for the right to dissent, as a threat to core values. Social and spatial boundaries in these circumstances become charged and energized. The defence of institutions like the family and spaces like the suburb becomes a more urgent undertaking during a moral panic. The oppressed, however, have their own strategies which challenge the domination of space by the majority, if only briefly and in prescribed locales. Ultimately, carnivalesque events confirm their subordination.

The problems that I have been discussing here concern, in part, territoriality, the defence of spaces and transgressions. Space is implicated in many cases of social exclusion, and in the next chapter I will try to identify in more detail the characteristics of exclusive social spaces and to relate these spaces to questions of power and social control.

NOTES

1. Edmund Leach, *Culture and Communication*, Cambridge University Press, Cambridge, 1976. Although Leach's penchant for A's and not-A's and his generally arid style is rather off-putting and suggests an attempt to be scientific in his approach, he was sensitive to cultural difference. The formal treatment is only a means of exposing some key issues.

2. ibid., p. 35.

3. I am conscious of the fact that homes often contain families, the members of which may have conflicting interests. I discuss the question of power relations and exclusion in the home in Chapter 6.

4. This example comes from Roderic Lawrence, *Housing, Dwellings and Homes*, Wiley, Chichester, 1987.

5. The problematic nature of the category 'adolescent' is discussed by Allison James, 'Learning to belong: the boundaries of adolescence', in Anthony Cohen (ed.), *Symbolizing Boundaries*, Manchester University Press, Manchester, 1986, pp. 155–170.

6. Nanette Davis and Bo Anderson, *Social Control: The production of deviance in the modern state*, Irvington, N.Y., 1983.

7. Durkheim's concept of mechanical and organic solidarity is discussed in detail in Anthony Giddens, *Capitalism and Modern Social Theory*, Cambridge University Press, Cambridge, 1971, pp. 75–79.

8. ibid., pp. 75–76.

9. This is an assertion made by a number of writers, but, as I suggest later in this chapter, this kind of broad-brush portrayal of American suburbia needs qualifying. Apart from academic studies of purified suburban communities, there have been a few fictional ones, like the film *Edward Scissorhands*, which portrays the reaction of a conformist Californian suburban community to a gothic 'other', making a similar point to Richard Sennett in *The Uses of Disorder*, Penguin, Harmondsworth, 1970.

10. Mary Douglas's ideas on purity and defilement are further developed in *Natural Symbols*, Barrie and Jenkins, London, 1970.

11. Mary Douglas, *Purity and Danger*, Routledge and Kegan Paul, London, 1966.

12. ibid., p. 41.

13. ibid., p. 56.

14. ibid., p. 55.

15. Jacob Neusner, *The Idea of Purity in Ancient Judaism*, E. J. Brill, Leiden, 1973. Emphasizing the strength and pervasiveness of purification rituals, Neusner observes that:

> The land is holy, therefore must be kept clean. It may be profaned by becoming unclean. The sources of uncleanness are varied and hardly cultic: certain animals, women after childbirth, skin ailments, mildew in the house, bodily discharges, especially the menses and seminal fluid, sexual misdeeds and the corpse.

16. S. Murray, 'Fuzzy sets and abominations', *Man*, 18, 1983, pp. 396–399.

17. In the preface to the first edition of *Natural Symbols*, Douglas argued that 'Each social environment sets limits to the possibilities of remoteness and nearness of other humans and limits the costs and rewards of group allegiance to and conformity to social categories.'

18. Sennett, op. cit. In this book, Sennett drew on the psychoanalytical theories of Erik Erikson in developing his thesis on purified identities and purified suburbs.

19. Constance Perin, *Belonging in America*, University of Wisconsin Press, Madison, 1988.

20. Mike Davis, *City of Quartz*, Verso Press, London, 1990.

21. These concerns emerged in a British study of people's ideal homes. Untouched wilderness in proximity to the home was severely frowned upon, presumably because it was not ordered or regulated nature. Distance from areas of racial tension was a predictable preference. *Designing and Selling Three and Four Bedroom Houses*, Research Associates Ltd, Stone, Staffordshire, 1988.

22. James Davis, *Fear, Myth and History: The Ranters and the historians*, Cambridge University Press, Cambridge, 1986.

23. This is documented by M. Baldasarre, *Trouble in Paradise: The suburban transformation of America*, Columbia University Press, New York, 1986.

24. Stanley Cohen and Laurie Taylor, *Escape Attempts*, Allen Lane, London, 1976.

25. Stanley Cohen, *Folk Devils and Moral Panics*, MacGibbon and Kee, London, 1972, p. 9.

26. Chris Philo writes about 'postmodern as object', which supposedly constitutes 'a complex (and seemingly quite chaotic) collision of *all manner of different objects* in the messy "collage" of contemporary people and places' (Paul Cloke, Chris Philo and David Sadler, *Approaching Human Geography*, Paul Chapman, London, 1991, p. 179). I am doubtful about this claim because I see little evidence of a loosening up of state controls. Both the state and capital impose order on societies through the regulation of space, as Mike Davis demonstrates graphically in the case of Los Angeles, or as the British government is attempting to do by

restricting the activities of minorities in the countryside (see Chapter 6). The transgressions of folk devils are a symptom of this power. While there are some examples of unfettered expressions of difference, others are a product of exclusionary practices, or difference is recast as deviance in the process of exclusion.

27. Davis, op. cit.

28. ibid., pp. 104–105.

29. ibid., p. 106.

30. New Age Travellers are a diverse group of semi-nomadic people, primarily of urban origin and given a common identity by their rejection of mainstream aspirations and a search for autonomy in a rural setting. Richard Lowe and William Shaw, *Travellers: Voices of the New Age nomads*, Fourth Estate, London, 1993, is a useful ethnographic study.

31. In this account, I draw on Simon Watney, *Policing Desire: Pornography, AIDS and the media*, Methuen, London, 1987.

32. ibid., p. 40.

33. ibid., pp. 40–41.

34. Cited by Susan Smith, 'Crime and the structure of social relations', *Transactions, Institute of British Geographers*, NS, 9 (4), 1984, 427–442.

35. ibid.

36. David Kunzle, 'World Upside Down: the iconography of a European broadsheet type', in Barbara Babcock (ed.), *The Reversible World: Symbolic inversion in art and society*, Cornell University Press, Ithaca, 1978, pp. 39–94.

37. I am grateful to Philip Jones of the School of Geography, Hull University, for this information.

38. Natalie Davis, 'Women on top: symbolic sexual inversion and political disorder in early modern Europe', in Babcock, op. cit., pp. 147–190.

39. Peter Stallybrass and Allon White, *The Politics and Poetics of Transgression*, Methuen, London, 1986, p. 20.

40. For an analysis, see Peter Jackson, 'Street life: the politics of Carnival', *Environment and Planning D: Society and Space*, 6, 1988, 213–227.

41. I discuss this legislation in relation to 'others' in the English countryside in Chapter 6.

42. Jean-Pierre Liègeois, *Gypsies: An illustrated history*, Al Saqi Books, London, 1986.

4

MAPPING THE PURE AND THE DEFILED

There is a history of imaginary geographies which cast minorities, 'imperfect' people, and a list of others who are seen to pose a threat to the dominant group in society as polluting bodies or folk devils who are then located 'elsewhere'. This 'elsewhere' might be nowhere, as when genocide or the moral transformation of a minority like prostitutes are advocated, or it might be some spatial periphery, like the edge of the world or the edge of the city. In constructing these geographies, the imagery discussed in Chapter 2 is drawn on to characterize both people and places, reflecting the desire of those who feel threatened to distance themselves from defiled people and defiled places. Thus, values associated with conformity or authoritarianism are expressed in maps which relegate others to places distant from the locales of the dominant majority. Images of others in the mind, in literature and other media may, however, inform practice such as the isolation of Gypsies on local authority sites in Britain or the exclusion of children from adult spaces. There may be important connections between these fantasies and the exercise of power. I will trace some of these ideas about the constitution of social space according to which some groups or peoples are deemed not to belong over a long historical time period in order to demonstrate their persistence. Portrayals of minorities as defiling and threatening have for long been used to order society internally and to demarcate the boundaries of society, beyond which lie those who do not belong. To demonstrate this point, I will make references both to political discourse in a number of historical periods and to some fictional narratives which mirror social practice. One informs the other.

The expansion of European empires and the development of the capitalist world economy required fitting dependent territories and dependent peoples into the cosmic order of the dominant powers. Beyond the spatial limits of

Plate 4.1 The 'Plinian' races – discrepant others beyond the edge of the civilized world (from Sebastian Munster's *Cosmographei*, Basle, 1550)

civilization, there were untamed people and untamed nature to be incorporated into the imperial system. Attitudes to people on these peripheries were ambivalent, however. While they were regarded with disgust or fear if they violated the space of the colonizers, they were also idealized and romanticized. Thus, Friedman notes that the ancient Greeks and Romans, like mediaeval European powers, saw themselves at the centre of the civilized world, and in their ordering of cultures and societies, the farther a group was from the centre the greater was its 'vice'.[1] Some cultural difference may have been tolerated but if, as Constance Perin suggests, another people's culture were considered to be too discrepant, it would be considered deviant, a 'vice', and generally judged in negative terms. Thus, on a global scale, a spatial and cultural boundary was drawn between civilization and various uncivilized, deviant 'others'. Again citing Friedman, Perin makes the interesting point that Aristotle's conception of the mean or average was effectively a moral judgement about levels of civilization. Being close to it was a mark of virtue but departure from the mean signalled vice. Thus, deviation in the statistical sense was also moral deviance and a device for conceptualizing the boundaries of society.[2]

This conception of civil society was echoed in mediaeval and early modern European cosmographies, which borrowed heavily from classical Greek sources. Thus, the 'edge' of civilization was marked by the presence of grotesque peoples, as, for example, in Sebastian Munster's *Cosmographei* (Plate 4.1). These people are not entirely different from the messengers of civilization in physical appearance, but they are 'imperfect' – physically deformed and/or black and at one with nature, in other words, not quite human by civilized, white European standards. This sort of characterization, as Stallybrass and White have observed, betrayed fears of being less than perfect on the part of the civilized. They suggest that the grotesque was not only the 'other' of the defining group or self but also 'a boundary phenomenon of hybridization or inmixing, in which self and other become enmeshed in an inclusive, hetereogeneous, dangerously unstable zone'.[3] In other words, those threatening people beyond the boundary represent the features of human existence from which the civilized have distanced themselves – close contact with nature, dirt, excrement, overt sexuality – but these same characteristics are exaggerated in portrayals of the uncivilized, which employ negative images of smell, colour and physical form. The world map, with civilization in the centre and the grotesque adorning the periphery, then expressed this desire for a literal distancing from the 'other'. More generally, space was used to establish a hierarchy which distinguished the civilized European from uncivilized native peoples. Thus, Mason describes paintings of Brazilian Indians by Eckhout, a seventeenth-century Dutch artist, in the following terms:

> Eckhout orders the Indians in accordance with a scheme that is centred on European canons. The wild dance and the (ethnographically inaccurate) portrayal of the 'Tapuya' as cannibals situate them on the outer ring of wildness. The more 'civilised' Tupi Indians occupy an intermediary position, marked for example by the fact that the Tupi man bears a European knife, and that the landscape behind the Tupi woman contains rows of cultivated palms and a colonial house. The 'civilised' Europeans come . . . in the centre as the most attractive and refined people.[4]

Although this kind of differentiation is dependent on disgust, the very features which are reviled are also desired because they represent those features of the civilized self which are repressed. Defiled peoples and places offer excitement.

Thus, in the early period of European exploration and the emergence of capitalist economies, there was an evident fascination with non-European cultures, but there were both moral and economic arguments for representing these cultures as less than human, a part of nature, or monstrous. The moral case is explained by Friedman:

> if the races were signs from God, the question then arose, what were they meant to signify? This problem lent itself particularly well to the exigetical techniques so familiar to us in bestiaries, spiritual encyclopaedias, and other homiletic works of the later middle ages. It produced a number of *moral* interpretations of the races that made them figures for various virtues and vices. The end result was a heightened treatment of alien peoples in didactic literature, exaggerating their unusual qualities so as to bring out their 'monstrousness' in both the older and the newer senses (the older sense being 'a disarrangement in the familiar order of existence').[5]

Perin suggests that the maps which expressed this cosmic order became a source of moral authority, comparable in importance to early Christian and classical authorities. This moral authority is evident in the Ortelius world atlas *Theatrum orbis terrarum*. On the title page, four female figures represent Europe, Asia, Africa and America.

> A sonnet by Gérard du Vivier explained the symbolism of the four figures. The lines dealing with America dwell on cannibalism, savagery and infinite treasure. To the map of Europe Ortelius appended a note proclaiming Europe's historic mission of world conquest, in the process of fulfilment by Spain and Portugal . . . Ortelius declared that the inhabitants of Europe had always surpassed all other peoples in intelligence and physical dexterity. These qualities naturally qualified the European to govern other parts of the world.[6]

The economic argument for monstrous representations, opposed to the perfection of white Europeans, was to ease the way for genocide in newly discovered territories, where, as Wallerstein has suggested, physical resources like gold were valued above a sustainable supply of labour by the colonizing powers during the early phase of capitalist development.[7] It could be argued that elements of the monstrous tradition have continued into the twentieth century in capitalist states, for example, myths about cannibalism among colonized peoples. 'Under the glaze of centuries of civilization are unexamined, if not mediaeval meanings and symbols which maintain and sustain our frightened apprehension of difference.'[8]

In mediaeval Europe, there is some evidence that the socio-spatial structure of the city also expressed a wish to erect boundaries to protect civil society from the defiled, even though some writers, notably Lewis Mumford,[9] have idealized mediaeval cities as socially integrated collectivities. My evidence comes from Geremek,[10] whose account of the social topography of Paris suggests that the bourgeoisie were scandalized by the behaviour of deviant groups and attempted to control their distribution. In particular, prostitution, although legal, was spatially regulated and 'red-light' districts were contested spaces, frequently objected to by respectable citizens.

> The basic principle of mediaeval regulation was to designate certain areas to prostitution either inside or outside the walls and limit vice strictly to them. The aim of this sort of social hygiene was to locate these places well away from the burgess or seigneural residences.
>
> (p. 87)

Some of the language of exclusion appears in Geremek's concluding comment on prostitution in the city, although it is not entirely clear in the following passage whether the terms are his or are taken from documentary sources:

> In spite of the ambiguities of mediaeval attitudes to prostitution, in spite of the elements of integration into the town that we have been able to show, in spite, finally, of the tolerance demonstrated in these matters by the law, the whole world inevitably declined towards marginality. As we have progressed along the 'streets of shame' of mediaeval Paris, among the 'shops of sin', we have constantly come up against people whose way of life, if not their moral code, place them outside the structures of society.
>
> (p. 87)

On Geremek's evidence, stereotypes of people and place were not as clearly articulated as they were in the capitalist city of the nineteenth and twentieth centuries partly because distancing, in a physical sense, was not easily accomplished in the compact and crowded mediaeval city. This was not for want of trying. Markus describes the city gates as a 'purifying filter', where strangers under arrest would be confined to prevent them contaminating civil society.[11] It is unlikely that this filter was effective, however, and inside the city the defiled were associated with a particular street or house rather than a district. However, some of the documents described by Geremek give a sense of bourgeois disgust at poverty, criminality and prostitution and at the places with which these conditions and activities were associated.

In the modern period in Europe, the language of defilement is more readily identifiable, as are the spaces to which are assigned those who belong and those who are excluded. By the eighteenth century, socio-spatial separation was becoming characteristic of large cities, like London, Dublin or Philadelphia, and boundary maintenance became a concern of the rich, who were anxious to protect themselves from disease and moral pollution. This is suggested somewhat obliquely in Swift's *Gulliver's Travels*, which can be read as a critique of western European society in the eighteenth century, using metaphors of purity and defilement. Swift creates a series of landscapes in which Gulliver is either polluting or is trying to protect himself from the threat of pollution. Drawing on Mary Douglas's work, Hinnant suggests that 'In every voyage undertaken by Gulliver, the impure is what escapes categories or threatens their existence: the unclean is the anomalous, the ambiguous or the monstrous.'[12] Thus, Lilliput is a highly ordered society with strong rules of exclusion where Gulliver, differing not only in size but also in behaviour, is polluting. Because he is a source of defilement here, Gulliver is consigned to a polluted space, the Temple:

> At the place where the carriage stopt, there stood an ancient Temple esteemed to be the largest in the whole Kingdom, which having been polluted some years before by an unnatural Murder, was according to the zeal of these People, looked upon as Prophane, and therefore had to be applied to common Uses and all Ornaments and Furniture carried away.

Notwithstanding the defilement of this space, because the pollution taboos in Lilliput were so strong, Gulliver was unclean and anomalous even here. As Hinnant observes:

> Among the Lilliputians, ethical pollution is measured in physical terms; their prohibition against bodily discharges extends even to structures which, like the temple, are no longer regarded as holy. Accepting this prohibition, Gulliver resolves henceforth to perform his natural functions as far beyond the precincts of the building as possible.
>
> (p. 18)

Consciousness of pollution in Lilliput is heightened by the geometry of the landscape. In particular, the metropolis, Mildendo, had a highly ordered design with strong internal boundaries and the populace was excluded from the centre – a sacred space, the home of the emperor. As in the European Baroque city on which Mildendo was probably modelled, geometry expresses power: the representation of the masses as polluting is a means of exercising control.

In Gulliver's second voyage to Brobdingnag, there is a reversal. Attitudes to social mixing are very relaxed and pollution taboos are not in evidence. Significantly, Hinnant recognizes (p. 31) that 'there is a total disregard for geometric figures', and this is symptomatic of the Brobdingnagians' integrated rather than segmented view of society. One telling feature of their socio-spatial organization is that, rather than maintaining hospitals for the incarceration of the old and diseased and others who are marginal or residual in Brobdingnagian society, 'They are willing to grant their beggars the liberty to roam freely through the streets of Lorbrulgrud.' The reversal in world-view represented by Brobdingnag in relation to Lilliput has interesting consequences for Gulliver.

In Lilliput, he is polluting because he is unable to conform, but in Brobdingnag, where, understandably, he fears for his survival because of his diminutive stature, Gulliver becomes preoccupied with boundaries. Hinnant (p. 33) notes that his reaction to Brobdingnagian society is 'visceral, embodied in the nausea he feels at the sight of promiscuous maids and beggars'. Thus, he insists on the separation of basic social categories – male and female, healthy and diseased, rich and poor – because mixing and non-conformity, like expressions of sexuality outside conventional bounds, create anxiety. Hinnant suggests that this concern for boundaries is essential if Gulliver is to survive

> in a country where he is subject to the predations of all but the most innocuous creatures. [The prohibitions] . . . seek to accomplish on the cultural plane what Gulliver inevitably fails to achieve on the plane of nature: to construct a shield that will protect him from threats of intrusion and destruction.

There is an interesting parallel in this with the small group on the margins of industrialized societies, like Gypsy communities, for whom pollution taboos and a concern with boundaries relate to the problem of cultural survival. *Gulliver's Travels*, however, can be seen more generally as a commentary on social tensions and power relations in a developing urban society as they are expressed in the language of defilement.

The poor as a source of pollution and moral danger were clearly identified in contemporary accounts of the nineteenth-century capitalist city. As socio-spatial segregation became yet more pronounced, the distance between the affluent and the poor ensured the persistence of stereotyped conceptions of the other. Social and spatial distancing contributed to the labelling of areas of poverty as deviant and threatening, a lack of knowledge being reflected in myths about working-class living conditions and behaviour. Dyos and Reeder convey nicely the kind of language which was used to describe working-class and bourgeois environments in a comment on class divisions in London: 'the undrained clay beneath the slums oozed with cesspits and sweated with fever; the gravelly heights of the suburbs were dotted with springs and bloomed with health'.[13] This expression of the class divide in terms of topography and health was crucial. The poor, down there on the swampy clays, were living in their own excrement and were subject to contagious diseases like cholera. The middle classes, up there on the suburban heights, were free from disease and uncontaminated by sewage, but threatened by the poor and their diseases. In one sense, the quotation identifies a serious public health problem which reflected rapid urbanization without provision of adequate services. In another sense, however, it is a comment on different standards of morality.[14] The poor were not only living in appalling physical circumstances but were, from a bourgeois perspective, *depraved*. Thus, as Stallybrass and White comment on the urban reformer Chadwick's attitude to the poor: '[He] connects slums to sewage, sewage to disease, and disease to moral degradation.' Their degradation was connected in Chadwick's view with their lack of control over desire: 'short-lived, improvident, reckless and intemperate, with an habitual avidity for *sensual* gratifications'. They continue:

> Like most of the sanitary reformers, Chadwick traces the metonymic associations (between the poor and animals, between the slum dweller and sewage) . . . But the metonymic associations, which trace the social articulation of depravity, are constantly elided with and displaced by a metaphoric language in which filth stands for the slum dweller: the poor are pigs.[15]

The significance of excrement in this account is that its stands for residual people and residual places. The middle classes have been able to distance themselves from their own residues, but in the poor they see bodily residues, animals closely associated with residual matter, and residual places coming together and threatening their own categorical scheme under which the pure and the defiled are distinguished. The separations which the middle classes have achieved in the suburb contrast with the mixing of people and polluting matter in the slum. This then becomes a judgement on the poor. The class boundary marked out in residential segregation echoes the recurrent theme: 'Evil . . . is embodied in excrement'.[16]

Similar moral judgements surface in some of Charles Booth's descriptions of London at the end of the nineteenth century.[17] While he demonstrates considerable sympathy for the working class, Booth still recognizes a residual population which puts itself beyond civil society through its behaviour and material circumstances – this group, an 'internal colonial other', is identified by black shading in his maps of social class. Thus, in a comment on Whitechapel in the East End of London, he contends that:

> There is a large class who must be regarded as outcasts, for whom the policy of sanitary regulation, of inspection, even of harrying, seems to be the only resource, and who must be regarded, in the mass, as hopeless subjects of reform.

However, Booth recognized that things were getting better. Thus,

> In spite of the wretched beings who sleep each night on the doorsteps in Commercial Street, and the worse figures which parade its pavements; in spite of the hells of Dorset Street, and the

low life and foul language of the courts; in spite of the poverty and drunkenness, domestic uncleanliness, ignorance and apathy, that still prevail – things are surely making for the better in Whitechapel and St. George's.

(pp. 64–65)

He adds that 'Such scenes of unmitigated savagery as old inhabitants have witnessed are unknown now.' Booth's social geography was a moralizing geography which linked the poor with dirt and deviant behaviour and defined particular residual spaces, both enclaves and exclaves. Typical of the latter was Notting Dale in north-west London, tellingly referred to as 'The Piggeries' and populated by Irish migrant labourers and Gypsies, among others. Documents like Booth's created powerful negative images which conditioned the response of the urban reformers. Physical cleaning, separating the poor from their residues, was to be accompanied by 'moral cleansing' or purification because 'moral filth was as much a concern as physical'.[18] Similarly, Corbin suggests that in Paris in the mid-nineteenth century 'The reformers nursed the plan of evacuating both sewage and vagrants, the stenches of rubbish and social infection all at the same time.'[19]

Nineteenth-century schemes to reshape the city could thus be seen as a process of purification, designed to exclude groups variously identified as polluting – the poor in general, the residual working class, racial minorities, prostitutes, and so on. This was particularly true of grand designs like that prepared by Haussmann for Paris. One of Haussmann's objectives, according to Knaebel,[20] was to make central Paris fit for the bourgeoisie by creating elegant spaces which distanced them from the poor and enhanced property values:

In Haussmann's eyes . . . there was a city to be embellished in those places where the bourgeois gave himself to the enjoyment of perception, where nothing must offend the senses – which implied the expulsion of the dirty, the poor, the unclean, the malodorous – and the 'non-city' [those who did not belong because they were not seen to be a part of civil society].

Ivan Illich also uses an olfactory metaphor to describe state interventions in the capitalist city in the nineteenth century, making a similar point to Knaebel about the distinction between the pure bourgeois and the defiled proletarian:

The effort to deodorize utopian city space should be seen as an aspect of the architectural effort to clear city space for the construction of a modern capital. It can be interpreted as the repression of smelly persons who unite their separate auras to create a smelly crowd of common folk. Their common aura must be dissolved to make way, to make space for, a new city through which clearly delineated individuals can circulate with unlimited freedom. For the nose, a city without aura is literally a 'Nowhere', a u-topia.[21]

These particular mappings of nineteenth-century urban society are not solely imaginery. There were chronic problems of sanitation, waste disposal and associated illnesses which urban reformers were intent on solving and, as progress made in methods of waste disposal weakened the association between the poor and excrement, so the bourgeois metaphors seemed less appropriate. Corbin suggests a number of stages in the separation of the pure self and the defiled other, a separation which was projected onto society and served to reinforce cleavages between social groups. The first separation was class based. Once the bourgeoisie developed a sense of self which excluded bodily residues, they could recognize their difference from the smelly working class:

Once all the smells of excreta had been got rid of, the personal odors of perspiration, which revealed the inner identity of the 'I', came to the fore . . . the bourgeois showed that he was increasingly sensitive to olfactory contact with the disturbing messages of intimate life.[22]

The abhorrence of excrement became an abhorrence of the poor, who represented what the bourgeoisie had left behind. Public health policies dealt with the problem of the putrid masses and cleaning up the poor would also help to instil ideas of discipline and order amongst them. Public health schemes brought with them regulations and were thus a means of social control. Thus, as Markus notes, the use of water in nineteenth-century European societies had a clear class dimension:

The apparently universal images of water on which Illich meditates – based on drinking, washing of bodies and clothes, germination, sport and health – became instruments for control in the baths, wash-houses and laundries for the poor . . . but metaphors for regeneration and visible ratification of superior status in the spas dedicated to the drinking of and bathing in mineral waters, and other elite baths.[23]

However, once the indigenous poor had been sanitized, Corbin argues, the same notions of dirt and disease could be used to construct images of immigrants, so defilement entered the language of racism.

I doubt whether there is a neat historical sequence here, but modern social geographies, that is, media and other popular representations of place rather than academic geographies, do suggest that spatial categories like 'the inner city' and some social categories, like Gypsies, are represented in similar language to that used to exclude the poor from bourgeois space in the nineteenth century. One crucial difference is that large sections of the working class are now more conscious of their own purified identity. In modern western societies, defilement is usually suggested in more muted tones than was the case in the nineteenth century. Material improvements in housing, water supply and sewage disposal have literally cleaned up the city, but I would agree with Corbin that places associated with ethnic and racial minorities, like the inner city, are still tainted and perceived as polluting in racist discourse, and place-related phobias are similarly evident in response to other minorities, like gays and the homeless.

Class-based geographies of defilement were still evident in the first half of the twentieth century, however, particularly where working people threatened the sanctity of middle-class preserves. Thus, middle-class commentaries on the countryside in Britain until the 1950s expressed concern about disorder, litter, advertising, and so on, associated with developments which were perceived to be catering for the working class, like roadside cafés and some housing developments. Plotlands, as working-class creations, were considered particularly abhorrent. Hardy and Ward cite one critic of Peacehaven, a plotland development in Sussex dating from the 1920s, who referred to 'the poison [beginning] at Peacehaven', and this kind of judgement was linked explicitly to class – 'the danger of proletarianism is near'.[24] Similarly, the architect Raymond Unwin, writing in 1929, averred that 'it is pathetic to see how often the exodus of those who find life in the modern town no longer tolerable destroys those very real amenities which they go forth to seek'.[25] Establishment figures like the Cambridge geographer, J. A. Steers, made very strong statements about working-class housing in the countryside, making it clear that it was considered to be a form of pollution. Steers described Canvey Island, on the south Essex coast, as 'an abomination . . . a town of shacks and rubbish . . . *It caters for a particular class of people* and, short of total destruction and a new start, little if anything can be done.'[26] Vocal objectors to spontaneous housing development in the countryside, like Unwin and Steers, had an important influence in shaping the legislation which formed the basis of town and country planning in England and Wales after 1945, and the power given to local authorities to control or eradicate 'disorderly development' under the 1948 Town and Country Planning Act contributed to the exclusion of working people from middle class space, particularly in areas of extensive plotland development, like Sussex and Essex. The rhetoric had an important bearing on practice, although the language of pollution was translated into less emotive terms, like non-conforming use.[27]

The same class prejudices of the period can be projected onto space in less explicit ways, for example, in the view of London contained in T. S. Eliot's *The Waste Land* (1922). This work is interesting not because it had any influence on policy, which would have been highly unlikely, but because, in conveying an elitist view of society, it also makes use of the language of defilement to describe socio-spatial relations. In *The Waste Land*, defilement occurs in the form of litter, corpses, fog – the residues which invade the social world of the bourgeoisie and disintegrate class boundaries. Discarded objects which were, according to Eliot, evidence of working-class recreation, intruded on bourgeois space. It was working-class pleasure and sexuality, in particular, which Eliot represented as sources of pollution, demonstrating exactly the same reaction as Unwin and Steers to the presence of the working class in rural England. These urban wastes threatened the stability and dominance of Eliot's class. Ellman argues that *The Waste Land* is essentially a poem about abjection, urban waste, both human and material, signalling the poet's anxieties.[28] However, the wastes are clearly working-class residues and, as she suggests, the filth which Eliot maps out in London insinuates defilement within.

MODERN MEDIA REPRESENTATIONS

Urban society, as it is currently projected in literature, film and television commercials, provides further visions of purity and pollution where the polluting are more likely to be social, and often spatially marginal minorities, like the gays, prostitutes and homeless mapped by Winchester and White.[29] Media representations are mostly fictional, imaginary constructions, but they draw on the same stereotyped images of people and places which surface in social conflicts involving mainstream communities and 'deviant' minorities. The media, particularly television, are also important because they comprise a major source of images for the representation of others, remotely consumed and requiring no engagement with the people they characterize as different. They are thus more likely to be received uncritically.

One graphic and probably not grossly exaggerated depiction of the pure and the defiled in the city is Martin Scorsese's *Taxi Driver.* This is a stark cinematic portrayal of prostitution in New York City, expressed largely in metaphors of defilement. The main character, Travis Bickle, expresses strong feelings of disgust and desire in relation to women. Thus, he is fascinated by pornography, but as he cruises the streets in the red-light districts in his taxi, he sees only 'filth'. His commentary on the city is all about dirt and the need to purify the spaces populated by prostitutes and the sexually deviant.

Travis writes in his diary:

May 10th . . . Thank God for the rain which has helped to wash away the garbage and trash from the sidewalks . . . All the animals come out at night – whores, skunk pussies, buggers, queens, fairies, dopers, junkies; sick, venal; some day, a real rain will come and wash all this scum off the streets.

And, similarly, when asked by a presidential hopeful, Palantine, for his view on what is wrong with the country, Travis volunteers this about New York City:

You should clean up this city here because this city here is like an open sewer, it's full of filth and scum and sometimes I can hardly take it. Whoever becomes the president should just really clean it up, you know what I mean. Sometimes I go out and I smell it. I get headaches, it's so bad, you know, they just like never go away, you know. It seems like the president should just clean up the whole mess here, should just flush it down the fucking toilet.

Against this background of defilement, Betsy, the woman Travis idolizes, personifies purity: 'She was wearing a white dress; she appeared like an angel out of this filthy mess; she is alone, they cannot touch her.' His own anxieties about dirt dominate the film, so when Iris, a child prostitute he hopes to save from the streets, suggests that she might go to a commune in Vermont, Travis feels uncomfortable. He says that places like that are dirty and he couldn't go to a place like that. His final purifying act is to destroy the pimps with extreme violence and this act of purification, notwithstanding the violence, makes him a local hero. While *Taxi Driver* is a film about a personal obsession, it could also be seen as a moral geography which has a wider currency. Consider, for example, a comment by a member of a right-wing gang in east Berlin:

Swen enthuses about the way a section of the Nationale Alternative [a neo-Nazi organization] brought 'order' to the railway station in Lichtenberg – another district of east Berlin. 'It used to be full of Romanians, drunks and tramps. Now they've cleaned everything away.'

(*Guardian*, 27 September 1991)

Similarly, in a letter to the *New York Post*, a resident of the East Village laments the decline of her neighbourhood:

Our cars and apartments are being burglarized by the street peddlers who sell their stolen bounty on Second Avenue and the adjoining streets; St. Mark's Place has become a haven for

Plate 4.2 'God bless the child'. The vulnerable child, a symbol of purity in a defiled city (from a Volkswagen television commercial, BMP DDB Needham. Photo: Charlotte Hicks)

pimps, prostitutes, drug dealers, head shops selling drug paraphernalia and assault weapons, illegal immigrants and a sundry collection of other undesirables. A neighborhood that was once the hub of multi-culturalism and neighborly pride has, over the past 20 years, become a cesspool of vermin.

(*New York Post*, 21 September 1994)

Travis, the young fascist and the resident of a deteriorating East Village express in strong terms attitudes towards people and place which are deeply embedded in western societies, although in liberal discourse they are conveyed with greater subtlety. These vivid social maps of the city are also used for navigational purposes by banks, insurance companies, the police and the social services, but their spatial demarcations become visible only through practices such as the withdrawal of financial services from 'high-risk' localities.

Subtlety is evident in some modern advertising which, while presenting more restrained comment on the 'other' than *Taxi Driver*, still presents urban society in oppositional terms, stressing the virtues of the pure by setting it against images of pollution. The purpose of this is to suggest the possibility of

Plate 4.3 The homeless as shadows, a residue in the urban landscape (copyright © 1990 Fleetway Editions Ltd)

achieving a comforting state of purity through consumption, a point which I will develop in the next chapter. For the moment, I want to demonstrate how television commercials and the modern media generally have used either city landscapes or urban sub-cultures to make distinctions between a positively valued inside and a threatening exterior world.

Some car commercials have made particular use of images of threat and danger to convey the idea of the car as a protective capsule which insulates the owner from the hazards of an outside world populated by various 'others'. One example is a Volkswagen commercial which made use of a young child (Plate 4.2) to symbolize purity in the defiled environment of New York City.[30] The commercial implies that the car will transport her securely through the city, to the safety of the suburbs or a commisionaired apartment building. The city's street people – homeless, mentally ill, drug addicts – are represented as remote but threatening, part of another world viewed from the safety of the Volkswagen. This image of a threatening but invisible 'other' appears similarly in Jonathan Raban's *Hunting Mr. Heartbreak*.[31] Raban was mistaken for a street person when deviantly sitting on a fire hydrant in Manhattan:

It was interesting to feel oneself being willed into non-existence by total strangers. I'd never felt the force of such frank contempt and all because I was sitting on a fire hydrant. Every one of those guys wanted to see me wiped out. I was a virus, a bad smell, a dirty smear that needed cleaning up.

The comic strip *The Shadows* presents a view of the future North American city in the same terms (Plate 4.3). In this case, the homeless have been totally dehumanized. They exist only as a residue.

Other geographies have been suggested in detergent commercials where, predictably, purification through cleaning, attaining a state of whiteness and virtue, is a continuing theme. The morality of cleanliness tends not to be so explicit as it was, for example, in the Health and Cleanliness Council's posters illustrated in Chapter 2, but the virtue of cleanliness can be suggested through associations of people and places. Thus, in a new, concentrated Persil commercial (1991), children are depicted as a part of 'the wild', untamed nature, which is their natural habitat but one which renders them uncivilized. Mother (Plate 4.4a) wonders how the children get so dirty at school. The children are then shown in an imaginary sequence tearing through the wilderness (Plate 4.4b), but with the boys doing more adventurous things than the girl. Their place in nature is confirmed by dirty clothes, face paint and headdresses (Plate 4.4c). The suggestion of an American Indian stereotype is interesting because this also locates the minority in nature rather than as a part of society. The children, however, are returned to society, cleaned and cared for by mother, with the help of Persil (Plate 4.4d), but civilization appears to be but a veneer as they set out again into the wilderness – for a geography lesson (Plate 4.4e). The children are portrayed as 'naturally' wild but it is clear that Persil is a civilizing influence, a necessary commodity in the suburban home, contributing to the creation of a purified environment in which children behave according to standards set by adults. The family home is the setting for a struggle against dirt and natural wildness. Consumption is encouraged by suggesting the undesirability of the soiled and polluted.

It is interesting to compare the representation of the child in the city in the Volkswagen commercial with these images of childhood conveyed by the Persil commercial. In the first, the child is pure and the city, or rather some of the stereotyped inhabitants of the city, constitute a threat to this purity. In the second, the children are defiled through their association with nature and purified by the civilizing influences of mother, home and detergent. Children can be simultaneously pure and defiled. Nature, likewise, can provide images of purity, often in contrast to the defiled city, or, as wilderness, it can be associated with people – children, indigenous minorities, and so on – in order to represent them as less than civilized and in need of purification.[32] These shifts in the use of images demonstrate the contradictions and ambiguities which characterize stereotypes and the complex associations of people and

(a)

(b)

(c)

(d)

(e)

Plates 4.4a–e (from top to bottom) The association of children with the 'wild' and the civilizing role of detergent, mother and home (Persil television commercial, copyright © J. Walter Thompson Company Ltd)

places which are used to map the spaces of the same and the other.

It could be argued that my selective use of media representations of purity and defilement exaggerates the extent to which this opposition figures in modern discourses. Historical accounts of the problem, particularly those of Corbin and Stallybrass and White, indicate that in social terms a concern with pollution has narrowed its focus to become a concern about racialized or non-conforming minorities rather than reflecting a bourgeois anxiety about the unwashed masses, most of whom have become a part of the purified majority. However, it may be that material progress and increased regulation of space by the state and private capital have contributed to a heightened consciousness of purity and defilement with a consequent increase in exclusionary pressures affecting, for example, children as well as conspicuously non-conforming minorities. There does not appear to be any consistent progression, no marginalization of dirt *en route* to a state of blissful shining whiteness.

Finally, I want to suggest how such imaginary geographies translate into

Plate 4.5 A residual population in a residual space – Gypsies camped under a motorway in Arles, south of France, 1990 (photo: author)

practice. The kinds of representations described here in literature and the visual media confirm stereotypes of people and places and inform attitudes to others. These attitudes assume significance in community conflicts and in the day-to-day routines of control. This is evident, for example, in the case of European Gypsies, to whom opposition is expressed in a consistent and highly predictable form. Here, the problem is that Gypsies' dependence on the residues of the dominant society, scrap metal in particular, and their need to occupy marginal spaces, like derelict land in cities, in order to avoid the control agencies and retain some degree of autonomy, confirm a popular association between Gypsies and dirt. The fact that Gypsies have strong pollution taboos and high standards of cleanliness, where there are adequate facilities for keeping their trailers or houses clean, is irrelevant. Because of their frequent association with residues and residual spaces, the perception of many *gaujes* (non-Gypsies) is that Gypsies are dirty. Consequently, the fear of 'polluting Gypsies' leads to attempts by the dominant society to consign them to residual spaces where the stereotypical associations are confirmed.[33] This is illustrated by Plate 4.5, where a group of French Gypsy families are camped in a marginal urban space, surrounded by rubble, relegated to an environment where ethnic identity and dirt are connected in a negative stereotype. The representation of social categories either side of a boundary defined by notions of purity and defilement and the mapping of this boundary onto particular places are not solely a question of fantasy. They translate into exclusionary practice.

CONCLUSION

The idea of society assumes some cohesion and conformity which create, and are threatened by, difference, although what constitutes a threatening difference has varied considerably over time and space. Nation-states may or may not claim to accommodate diversity but at the local level social and cultural mixing is frequently resisted. What I have suggested in this chapter is that there are enduring images of 'other' people and 'other' places which are combined in the construction of geographies of belonging and exclusion, from the global to the local. Historically, at least within European capitalist societies, it is evident that the boundary of 'society' has shifted, embracing more of the population, with the class divide in particular becoming more elusive as a boundary marker. The imagery of defilement, which locates people on the margins or in residual spaces and social categories, is now more likely to be applied to 'imperfect people', to use Constance Perin's term – a list of 'others' including the mentally disabled, the homeless, prostitutes, and some racialized minorities. Clearly, the labels which signal rejection are challenged and there is always the hope that, through political action, the humanity of the rejected will be recognized and the images of defilement discarded. There is no clear picture of progress, however. Feelings of insecurity about territory, status and power where material rewards are unevenly distributed and continually shifting over space encourage boundary erection and the rejection of threatening difference. The nature of that difference varies, but the imagery employed in the construction of geographies of exclusion is remarkably constant.

NOTES

1. John Friedman, *The Monstrous Races in Medieval Art and Thought*, Harvard University Press, Cambridge, Mass., 1981, p. 35 (cited by Constance Perin, *Belonging in America*, University of Wisconsin Press, Madison, Wis., 1988).

2. Perin, op. cit., pp. 146–151. It is interesting to note the later quality of statistical science as a moralizing discipline. Particularly in the work of the nineteenth-century statistician Francis

Galton, ranking procedures were used to provide a supposedly objective account of racial difference. His classifications were spurious and racist, a clear case of 'moralizing difference, to use Constance Perin's term. See C. Blacker, *Eugenics: Galton and after*, Duckworth, London, 1952, on Galton.

3. Peter Stallybrass and Allon White, *The Politics and Poetics of Transgression*, Methuen, London, 1986, p. 193.

4. Peter Mason, *Deconstructing America: Representations of the other*, Routledge, London, 1990, pp. 21–22.

5. Friedman, op. cit., p. 109.

6. Benjamin Keen, *The Aztec Image in Western Thought*, Rutgers University Press, New Brunswick, N.J., 1971, p. 156.

7. Immanuel Wallerstein, *Historical Capitalism*, Verso, London, 1983.

8. Perin, op. cit., p. 174.

9. Lewis Mumford, *The City in History*, Secker and Warburg, London, 1961.

10. Bronislaw Geremek, *The Margins of Society in late-Mediaeval Paris*, Cambridge University Press, Cambridge, 1987.

11. Thomas Markus, *Buildings and Power: Freedom and control in the origin of modern building types*, Routledge, London, 1993, p. 118.

12. Charles Hinnant, *Purity and Defilement in Gulliver's Travels*, Macmillan, Basingstoke, 1987.

13. Stallybrass and White, op. cit., pp. 127–128.

14. Felix Driver, 'Moral geographies: social science and the urban environment in mid-nineteenth century England', *Transactions, Institute of British Geographers,* NS, 13 (4), 1988, 275–287.

15. Stallybrass and White, op. cit., p. 131.

16. Perin, op. cit., p. 178.

17. Charles Booth, *Life and Labour of the London Poor,* 3rd Series, Religious Influences, Macmillan, London, 1902.

18. Driver, op. cit.

19. Alain Corbin, *The Fragant and the Foul: Odor and the French social imagination*, Harvard University Press, Cambridge, Mass., 1986.

20. Cited by Corbin, ibid., p. 268n.

21. Ivan Illich, *H_2O and the Waters of Forgetfulness*, Dallas Institute of Humanities and Culture, Dallas, 1984, p. 53.

22. Corbin, op. cit., p. 143.

23. Markus, op. cit., p. 146.

24. Dennis Hardy and Colin Ward, *Arcadia for All: The legacy of a makeshift landscape*, Mansell, London, 1984.

25. Raymond Unwin, Greater London Regional Planning Committee, 1929, p. 27.

26. Hardy and Ward, op. cit., p. 120.

27. Le Corbusier had a similar modernist vision according to which small-scale, 'disordered' development was a form of pollution.

His plan for the redevelopment of the Right Bank in Paris (1925) showed that he did not care for people, for their bustle, traffic and markets. He proposed to replace the genial disorder of Rue de Rivoli, Les Halles and the Faubourg St. Honoré with a grid of cruciform tower blocks. He argued: 'Imagine all this junk, which has until now lain spread out over the soil like a dry crust, cleaned off and carted away and replaced by immense crystals of glass.'

(Centipede, *Guardian*, 18 March 1993)

28. Maud Ellman, 'Eliot's abjection', in J. Fletcher and A. Benjamin (eds), *Abjection, Melancholia and Love: The work of Julia Kristeva*, Routledge, London, 1990, pp. 178–200.

29. Hilary Winchester and Paul White, 'The location of marginalized groups in the inner city', *Environment and Planning D: Society and Space*, 6, 1988, 37–54.

30. The use of the child, particularly white girls, as a symbol of purity but, ambiguously, as a sexualized, defiled, object is discussed by Sander Gilman, *Sexuality: An illustrated history*, Wiley, Chichester, 1989, pp. 271–273, and by Rex and Wendy Stainton Rogers, *Stories of Childhood: Shifting agendas of child concern*, Harvester Wheatsheaf, Hemel Hempstead, 1992, pp. 181–187. In Graham Ovenden's *Victorian Children*, Academy Editions, London, 1971, photographic portrayals of the pure and defiled (child prostitutes) are juxtaposed rather starkly. In modern representations of the child, in advertising, for example, purity generally seems to be emphasized in order to underline the need for protection from a dangerous social and physical environment, as the Volkswagen commercial implies.

31. Jonathan Raban, *Hunting Mr. Heartbreak*, Collins Harvill, London, 1990.

32. The deceit involved in representing nature as pure is suggested by John Law and John Whittaker in a discussion of the depiction of sacred (pure) and profane (polluted) landscapes in literature on the acid rain problem:

nature is simplified and it is represented as pure. This is not so easy, for those who actually venture into the wilderness know that it is full of biting flies, carcasses, dead trees and land-slips, and they also know that power lines, roads and quarries abound. The pictures have thus been carefully selected and framed, for the production of a sacred representation of nature requires a technology of purification.

(John Law and John Whittaker, 'On the art of representation: notes on the politics of visualisation', in Gordon Fyfe and John Law (eds), *Picturing Power: Visual depictions and social relations*, Sociological Review Monograph 35, Routledge, London, 1988, p. 173)

33. David Sibley, 'Outsiders in society and space', in Kay Anderson and Fay Gale (eds), *Inventing Places: Studies in cultural geography*, Longman Cheshire, Melbourne, 1992, pp. 107–122.

5

BOUNDING SPACE: PURIFICATION AND CONTROL

> To be sure a certain theoretical desanctification of space (the one signalled by Galileo's work) has occurred, but we may not have reached the point of a practical desanctification of space. And perhaps our life is still governed by a certain number of oppositions that remain inviolable, that our institutions and practices have not dared to break down. These are oppositions that we regard as simple givens: for example, between private space and public space, between family space and social space, between cultural space and useful space, between the space of leisure and that of work. All these are still nurtured by the hidden presence of the sacred.
>
> (Michel Foucault)[1]

So far in this discussion, space has been hovering on the margins. I will now suggest that, in order to understand the problem of exclusion in modern society, we need a cultural reading of space, what we might term an 'anthropology of space' which emphasizes the rituals of spatial organization. We need to see the sacred which is embodied in spatial boundaries. In the quotation above, Foucault implies that a desanctification of space is occurring in western societies. This lags behind the desanctification of time, he suggests, but is an inevitable consequence of modernization, the progress of materialism and rationality. I doubt that this is the case. There seems to me to be a continuing need for ritual practices to maintain the sanctity of space in a secular society. These rituals, as in ancient Israel or Brobdingnag, are an expression of power relations: they are concerned with domination. Today, however, the guardians of sacred spaces are more likely to be security guards, parents or judges than priests. They are policing the spaces of commerce, public institutions and the home rather than the temple.

In Chapter 3, I indicated that there were parallels between social behaviour in small, high-density collectivities generally described as traditional societies and behaviour in present-day *gemeinschaft*-like groups. The liminal discourse of social anthropology developed in the context of traditional societies, concerning boundary rituals, taboos, and so on, can be used to illuminate the modern problem, although I would not advocate an exclusive disciplinary perspective. In the following account, however, I will try to identify the 'curious rituals' associated with the social use of space in developed societies, inter-leaving broad theoretical concepts and particular details of individual and group behaviour. Before looking specifically at the interconnections of spatial structures and social exclusion, however, we might consider a few general issues involved in unravelling socio-spatial relationships.

STRUCTURATION THEORY AND SPATIAL THEORY

While, in the history of modern geography, the nature of the relationship between people and the environment has been one of the more enduring concerns of practitioners, interest in the question faded in the 1960s when space was reduced to the primitives of distance and direction and served essentially as a neutral medium for the operation of social and economic processes. Foucault's observation about the treatment of space in the western philosophical tradition seems particularly apposite as a comment on the treatment of space in human geography: 'Space was treated as the dead, the fixed, the undialectical, the immobile. Time, on the contrary, was richness, fecundity, life, dialectic.'[2] Subsequently, an interest in structure in the materialist sense has led to a revived interest in environment, particularly in the built environment as a product of capitalist development. Conceptions of the way in which the environment affects and is affected by human activity have been presented by several writers recently, including Allan Pred[3] and Ed Soja,[4] who draw on Anthony Giddens's structuration theory. Pred, for example, asserted that

> Place . . . always involves an appropriation and transformation of space and nature that is inseparable from the reproduction and transformation of society in time and space. As such, place is characterized by the uninterrupted flux of human practice – and experience thereof – in time and space.

This sounds impressive although the writer is not saying anything particularly remarkable. The problem is that geographers in their earlier grossly simplified spatial geometries had neglected the obvious.

Giddens's account of structure and agency in the constitution of social life provides one point of entry to this problem. Although his structuration theory is now treated as rather passé in human geography, some of his ideas are useful in the sense that his conception of structuration provides cues for the unravelling of socio-spatial relationships. While Giddens seems to me to have a rather naive view of space, working with a few key arguments from his general thesis, we can begin to give shape to a socio-spatial theory of exclusion.

Giddens's theory of structuration is concerned with social relationships which are both fluid and concrete, and it is an argument which can be readily spatialized.[5] His first proposition is that human activities are *recursive*, that is, 'continually recreated by [social actors] by the very means whereby they express themselves as actors'. Second, the reproduction of social life presumes *reflexivity* in the sense that 'the ongoing flow of social life is continually monitored'. The monitoring of social life, however, also includes the monitoring of the physical contexts and the broader social contexts of experience. These contexts have structural properties which are 'both medium and outcome of the practices they recursively organize'. As this suggests, structure does not just constrain activity but is also enabling, although the agency of actors, their capacity to affect the circumstances of their existence, will not be equal in relation to all the structured properties of the social system. Some of these structured properties 'stretch away in time and space, beyond the control of individual actors'.

In addition to location, which, as Giddens implies, embodies a set of structuring spatial and temporal relationships, we can recognize the built environment as a relatively stable element of the socially produced environment which provides the context for action. Here, the reciprocity of human activity and its context is fairly obvious. As Arthur Miller said about society, 'The fish is in the water and the water is in the fish.' This observation, banal as it is, captures a characteristic of the built environment which is still neglected in much urban geography, however, with space represented too often as an inactive context for something else, the 'where' in a Kantian tradition, dead space. Giddens himself is not very clear on this question. At one point, he implies that spatial structures serve only as containers for social interaction. Thus, in a passage selected by Nicky Gregson:

Locales refer to the use of space *to provide settings for interaction* . . . Locales may range from a room in a house, a street corner, the shop floor of a factory, towns and cities, to the territorially demarcated areas occupied by nation-states. But locales are typically internally regionalized and the regions within them are of critical importance in constituting the contexts for interaction [my italics].[6]

As Gregson notes, this says nothing about 'the nature, form or content of the setting', or, I would add, about the interaction of people and the built environment. Later in the same work, however, Giddens claims that 'space is not an empty dimension along which social groupings become structured, but has to be considered in terms of its involvement in the constitution of systems of interaction'.[7] It is this assertion which is echoed in geographical accounts of structuration theory, by Pred and Gregson. It is, then, important to contextualize structuration theory, to recognize how particular social and spatial outcomes are tied to particular cultures, to particular histories and to individual life experiences.

While structuration theory points to the reciprocal nature of the relationship between people, as individuals and social groups, and their environment, it still leaves a problem of explanation, which has been identified by Steve Pile.[8] That is:

after the division of 'the social' into structure and agency (or into context and intentionality), structure (context) is seen as external while agency (intentionality) is seen as internal. The effect of externalizing structure is to make it taken-for-granted (not yet known) and impersonal (denying the personal in the social).

However, as I suggested in my earlier outline of object relations theory, structure is internalized (through introjection) and shaped by the unconscious (through projection). Again, to quote Pile:

Psychoanalytic theory, in its theories of the unconscious, describes how the social enters, constitutes and positions the individual. Similarly, by showing that desire, fantasy and meaning are a (real) part of everyday life, it shows how the social is entered, constituted and positioned by individuals.

I do not think that this explanatory gap in structuration theory makes it necessary to abandon it, and it has particular value in defining the problem of power. Recognizing that people have a capacity to change their environment and, more generally, that individuals retain some autonomy as thinking and acting agents, leads to the question of the distribution of power within social systems and of spatial structures as embodiments of power relations. As Moos and Dear observe:

> Power relations are always relations of autonomy and dependence and are necessarily reciprocal. The distribution of power in a relationship may be very assymetrical but an agent always maintains some control in the relationship and may escape complete subjugation.[9]

Control by dominating agents may seem complete, but there is always the possibility of subversion. The prison, as possibly the most dominating control environment, demonstrates the existence of autonomy in the most adverse conditions. Michael Ignatieff,[10] for example, notes that in Pentonville prison in London, designed in the 1840s as a model of the total institution, the impossibility of total control was recognized after a few years of a very harsh regime of hard labour and solitary confinement. The resistance of the prisoners led to a moderation of the system. There was, then, more than a flicker of human agency which altered the relationship between the institutional environment and the inmates. We cannot understand the role of space in the reproduction of social relations without recognizing that the relatively powerless still have enough power to 'carve out spaces of control' in respect of their day-to-day lives.[11]

We can envision the built environment as an integral element in the production of social life, conditioning activities and creating opportunities according to the distribution of power in the socio-spatial system. For some, the built environment is to be maintained and reproduced in its existing form if it embodies social values which individuals or groups have both the power and the capacity to retain. For others, the built environment constitutes a landscape of domination. It is alienating, and action on the part of the relatively powerless will register in the dominant vocabulary as deviance, threat or subversion. This contrast suggests that power relations are transparent, however, when they are not. In the routines of daily life, most people are not conscious of domination and the socio-spatial system is reproduced with little challenge. There are some groups for whom exclusion is a part of their daily experience, who will be highly sensitive to alien environments, but their spaces of control are too small to interrupt the reproduction of socio-spatial relations in the interest of the hegemonic power.

An appreciation of power relations gives meaning to space. Variations in the control and manipulation of different spatial configurations reflect different forms of power relations. As Foucault maintains,

A whole history remains to be written of *spaces* – which would at the same time be the history of powers (both of these terms in the plural) – from the great strategies of geopolitics to the little tactics of the habitat.[12]

The range of spaces which should be of interest to the human geographer interested in power relations is somewhat wider than that which has conventionally constituted the geographer's terrain, however. In particular, I will suggest that personal space defined by the self and the intimate spaces of the home are integral elements of social space. These private spaces have a relationship with the public spaces of geography – they are reciprocally conditioned, and it is the process of reciprocal conditioning which requires illumination if we are to understand problems like the rejection of difference in localities.

EXCLUSIONARY SPACE

I will argue that 'spatial purification' is a key feature in the organization of social space. Michel de Certeau recognized this problem as the creation of 'clean space' in Utopian and urbanistic discourse. He argued that:

In this site [the city] organized by 'speculative' and classifying operations, management combines with *elimination*: on the one hand, we have the differentiation and redistribution of the parts and function of the city through inversions, movements, accumulations, etc., and, on the other hand, we have the rejection of whatever is not treatable and that, thus, constitutes *the garbage of a functionalist administration* (abnormality, deviance, sickness, death, etc.) [my italics].[13]

He continues with an observation that is close to Stanley Cohen's view of social control which I discuss later in this chapter, namely, that 'progress, of course, allows for the reintroduction of an increasing proportion of the wastes into the management network and the transformation of those very flaws . . . into means for strengthening the system of order'.

This argument, which clearly resonates with the notion of abjection and pollution, needs to be given a more explicit economic dimension. We can see that the imperative of accumulation under capitalism has made developed societies centres of consumption within the global economy, and the way in which consumption is promoted, the process of 'want creation' identified by Galbraith, contributes to purified identities and feelings of abjection in relation to the 'other'. Fred Hirsch argued that the market economy in developed societies encouraged 'the strengthening of self-regarding individual objectives', and consumer advertising, he suggested, comprised 'a persistent series of invitations and imperatives to the individual to look after himself [*sic*] and his immediate family'.[14] Thus, the never-ending invitations to consume further the privatization of the family, which is closed off from the outside world. Life beyond the home enters the private sphere through stereotyped images, conveyed by videos, television commercials and similar media messages. Within the private world of the home, advertisers foster a negative view of soiled goods and a positive view of new, completely packaged domestic environments, clearly in order to maintain the levels of demand for domestic products. The imagery of this advertising is significant. It often promotes cleanliness, purity, whiteness and spatial order, images reflecting the idea of a pure inner self as, for example, in the Persil commercial described in Chapter 4, features which Freud associated with civilization and the sublimation of instinctual feelings. Unsullied whiteness is also associated with a germ-free environment so that a concern with maintaining a state of pure whiteness becomes a virtue – mother (usually) has to fight germs by keeping the house clean in order to protect her children, notwithstanding the fact that some of these 'germs' are necessary for health. Thus, the consciousness of dirt and disorder is increased and we can anticipate that a feeling of abjection will be particularly strong in those environments, domestic interiors, neighbourhoods which are symbolically pure. It is the identification of numerous residues, to be expelled from the body, the home and the locality, which is characteristic of this purification process. In such environments, difference will register as deviance, a source of threat to be kept out through the erection of strong boundaries, or expelled.

THE FORM OF PURIFIED SPACE

The anatomy of the purified environment is an expression of the values associated with strong feelings of abjection, a heightened consciousness of difference and, thus, a fear of mixing or the disintegration of boundaries. This is one of several possible maps of social organization which we can construct, drawing particularly on schemata developed by Basil Bernstein. Bernstein's project was concerned with control in educational systems, but his ideas have a particular resonance in relation to the question of exclusion. He recognized an affinity with Mary Douglas, who approached what was essentially the same problem through an analysis of the rituals surrounding purity and defilement, and, recently, similarities have been noted in the writing of Mary Douglas and Julia Kristeva.[15]

Bernstein's interest in educational sociology has been primarily in language and the curriculum, where he has produced a classification which links academic subjects (or other social objects) with modes of control, and it is this scheme which links the social and the spatial. It provides us with a means of identifying exclusionary structures. His general thesis shows the influence of Durkheim, particularly the latter's distinction between mechanical and organic solidarity. Thus, according to Atkinson,[16] Bernstein characterizes a social organization displaying mechanical solidarity as one which is segmented: 'members are arranged in relatively insulated, self-contained units' and 'roles are . . . ascribed in terms of a small number of primitive categories'. Conversely, organic solidarity is expressed through increasing individualization and a 'weakening of boundaries which formerly defined structural segments'. Bernstein does not use the mechanical–organic dichotomy in a temporal sense, signalling a change from traditional to modern, however. The terms are used instead to indicate different forms of organization within modern institutions. Mechanical solidarity, like *gemeinschaft*, is a characteristic social form in developed societies.

The particular form of the mechanical and the organic are presented in an educational context. Here, Bernstein represents the school curriculum as a number of subject areas insulated from each other in different degrees, according to the prevailing ideology. First, in 'Open schools, open society?' he distinguishes two polar types of curriculum organization which have the characteristics shown in Table 5.1.[17]

Table 5.1 Characteristics of open and closed curriculum organization

	Open	*Closed*
1	Ritual order celebrates participation and cooperation	Ritual order celebrates hierarchy and dominance
2	Boundary relationships with outside blurred	Boundary relationships with outside sharply drawn
3	Opportunities for self-government	Very limited opportunities for self-government
4	Mixing of categories	Purity of categories

In a later paper,[18] he rephrases the problem, using the terms *classification* and *framing* to describe the characteristics of mixing or purification in curricula. With strong classification, the contents of subject areas are strongly bounded and kept separate, while strong framing suggests a clear distinction between what may and may not be transmitted within subjects. Decisions on what is permissible come from above and inter-subject communication is minimized. Conversely, with weak classification, subject boundaries are weakly defined and there is less concern with the singular and distinct identities of subjects, and weak framing allows the transmission of a wide range of ideas within a subject. Strong classification and strong framing tend to go together, as do weak classification and weak framing, although alternative combinations are possible. When the curriculum is strongly classified, new ideas on pedagogy or academic content are seen to be threatening because they challenge the hierarchical control structure. A weakly classified system, by contrast, is a tolerant one in which new ideas are absorbed. They do not threaten non-hierarchical power relationships precisely because power is diffuse. A hierarchical power structure does not like ambiguity. Fuzzy boundaries between subjects in the school curriculum, for example, suggest communication between subjects which could represent a challenge to those in power. Therefore, some knowledge within a strongly classified system would be seen as 'dangerous knowledge', to be suppressed, ignored or rejected, if it did not fit the classification.[19] This is characteristic of polluting objects and ideas in Mary Douglas's thesis. They do not fit a society's classificatory system.

Bernstein's educational model provides a clear analogue for the structuring of social space. Thus, we can speak of strongly classified space, where there is internal homogeneity and clear, strong boundaries separate that space from others. Alternatively, we could identify a strongly classified spatial system, consisting of a collection of clearly bounded and homogeneous units, like land-uses in a city or the rooms in a house. The contents and arrangement of the contents of strongly classified space, like the furniture in a room, would be strongly framed if there were inflexible rules determining those internal arrangements. Difference in a strongly classified and strongly framed assemblage would be seen as deviance and a threat to the power structure. In order to minimize or to counter threat, the threat of pollution, spatial boundaries would be strong and there would be a consciousness of boundaries and spatial order. In other words, the strongly classified environment is one where abjection is most likely to be experienced. Strong classification will reinforce feelings of abjection and the two may be recursively related. Weak classification and framing as forms of spatial structure would be associated with social mixing, a tolerance of difference and little interest in boundary maintenance. It is also possible, but less likely, that strong classification will be combined with weak framing. Alaszewski,[20] for example, drawing on Douglas rather than Bernstein, describes fluid and relaxed regimes, incorporating a wide range of therapies, in some of the wards of a mental hospital. The hospital as a structure and in its institutional organization is strongly classified but, within it, there are instances of weak framing. Bernstein is concerned with polar types and in practice we might expect some problems in classifying environments and forms of social organization which do not match his model. He does provide us, however, with a basis for connecting social structures and spatial structures and, at the same time, we can make his model relate to psychoanalytical, anthropological and economic theory at the point where Freud, Kristeva, Sennett, Douglas and Hirsch converge.

SPACE AND SOCIAL CONTROL

Bernstein provides a link between exclusionary processes which are rooted in family and group relationships and exclusion which has its source in institutional practices. Classification and boundary maintenance are characteristic of both, and families, communities and institutions are all implicated in the construction of deviance and the exclusion of deviant individuals and groups. For the moment, however, I want to examine socio-spatial exclusion as a part of the more general question of social control, with particular emphasis on controls exercised by agencies of the state. Social control is a term which has varied usage, but what I will be concerned with here is the attempted regulation of the behaviour of individuals and groups by other individuals or groups in dominant positions. Specifically, I am concerned with constraints on social interactions and the use of space which result from the actions of social control agencies.

Davis and Anderson suggest a scheme for classifying social control systems, containing several elements which are relevant to the social control problem (Table 5.2).[21]

In this scheme, they distinguish between those controls which are external in origin and which are transmitted hierarchically and those which are

Table 5.2 A classification of social control systems

Mode of control	*High pervasiveness*	*Low pervasiveness*
External	Asylums, prisons	Bureaucracies, firms
Internalized norms and values	Transformative groups, for example, Alcoholics Anonymous	Self-help groups, for example, Weight Watchers
External and internalized norms and values	Traditional families and kin groups	Professions (law, medicine, etc.)

internalized, in that members of a group make a commitment to norms and values. The dimension of pervasiveness separates the total institution with a control or corrective function, like a prison, and the conformist community, on one hand, and organizations where control is an unstated objective, on the other. The externally controlled/highly pervasive category is the one which has most immediate relevance to the problem of exclusion. In Bernstein's terms, we are concerned here with hierarchy, strong classification and, by implication, a high level of visibility for those identified as deviant. However, while this scheme has heuristic value, control regimes should not be thought of as fitting into discrete categories. Thus, the asylum and the prison, rather than being considered exceptional, should be thought of as models which have a wider application in society even though they may assume a more muted form. In particular, pervasiveness should be thought of as a continuum rather than a dichotomous variable. This is the essence of Foucault's argument in *Discipline and Punish*, a text that has generated considerable discussion in the social sciences, including geography.[22]

Foucault's thesis is that the discipline of a highly controlled institution like a prison or a nineteenth-century asylum 'represents a continuation and intensification of what goes on in more ordinary places',[23] and that the controls which are embedded in ordinary life legitimate the kind of regime practised in a prison, for example. 'All micro-forms of discipline are functional to a larger system', as Michael Walzer puts it. Foucault's particular vision of a controlled society originates in Bentham's panopticon, which was a model for a totally controlled institution, designed on the principles of discipline, surveillance and hierarchical classification. The panopticon was a prison/ factory, so designed that the controller could remain invisible and at a distance from the inmates yet control their lives in detail. The panoptic principle, however, extends much more widely than this. As a metaphor for control, the panopticon 'inserted the power to punish much more deeply into the social body'.[24] It 'colonizes' social life and erects boundaries between the normal and the deviant at all levels, irrespective of legal codes which define criminal behaviour. Thus, control, discipline and carceral forms of punishment are diffused through society and social control on the panopticon principle becomes much more than confinement under a particular regime: 'The prison is only one small part of a highly articulated, mutually reinforcing carceral continuum extending across society in which all of us are implicated, and not only as captives and victims'.[25]

This is a bold claim and, as an account of the geography of social control, it warrants critical examination. The first problem concerns the generality of the panoptic principle. Prisons, asylums and workhouses, associated particularly but not uniquely with the disciplining of the proletariat in the nineteenth century, could be seen as useful instruments for the spatial exclusion of deviant and unproductive groups at a time when the Benthamite principle of getting the maximum return from labour encompassed the factory and the institution. Thus, institutions like the prisons or the magdalens, hostels for the confinement of 'prostitutes', were places for closely supervised work as well as for the correction of deviance. The geography or spatial design of these institutions varied. They were not necessarily planned on the panoptic principle, although they did generally exhibit strong classification. Thus, in his detailed account of buildings designed for 'formation', like schools, and for 're-formation', like workhouses, in capitalist societies during the eighteenth and nineteenth centuries, Thomas Markus argues that:

> In all these places, order is based on stable categories of people, objects and activities, together with a set of rules – much stronger and more explicit than in other buildings – which govern their interactions. They establish diurnal, weekly, and seasonal timetables and shifts, and they specify the duration and repetition of events. The rules are, equally strongly, built into space and its management. They define the location of persons and things, they control the paths of movement and the degree of choice as well as the visual paths, they define programmed encounters and place limits on those occurring by chance. Time and space are joined in rules which govern the opening times of specific spaces. In short, the building and its management determine who does what, where, with whom, when and observed by whom.[26]

We can also see the finely meshed network of control, represented physically in the design of the prison or the school, extending to other locales and to other social groups who might interfere with the efficient performance of the capitalist economy. This is particularly evident in specialized spaces, those which, like institutions, are based on an explicit ideology. Thus, the Utopian creations of nineteenth-century capital, like Robert Owen's New Lanark or Titus Salt's Saltaire, extended the discipline of the workplace to the residential sphere in that tenants were selected for their respectability and conformity to the ideals of the community. In the highly ordered space of the Utopian settlement, deviance would have been conspicuous. A more recent example is the settlements provided by the state for minorities whose presence interferes with the exploitation of resources by capital and whose values are in conflict with the materialistic, progressive values of capitalism. Planned settlements for Australian Aborigines, native Canadians and some European travelling people, for example, express the state's interest in separation and the correction of deviance.[27] Locations are selected which remove the minority from areas valued by the dominant society and, in isolation, the design and regulation of space are supposed to induce conformity. The regularity of the design, the high visibility of internal boundaries which interrupt traditional patterns of social organization, make what is culturally different appear disruptive and deviant. As in prison, power and domination are expressed in arbitrary rules and transgression warrants the imposition of sanctions, including eviction in the case of many English Gypsy sites. These schemes fail because there is no awareness of the capacity of the minority to resist and to maintain its own cultural values, but they do demonstrate the need of the state to secure the interests of capital through socio-spatial control of 'deviance' and cultural difference.

The mentally ill and mentally disabled, the criminal and the racially different are all in varying degrees 'other' and beyond the bounds of normal society according to some narrower definitions of normality. Do we, however, create spaces for the disciplining of groups within mainstream society, extending the finely meshed network of domination into areas of social life occupied by the majority? Rather than thinking of the problem as one of inserting panoptical controls into the social body, I think we should recognize the reciprocal conditioning of individuals and families, on the one hand, and social institutions on the other. Object relations theory, as I indicated above, suggests that the tendency to reject difference and to value order is characteristic of the pathological personality but that this tendency is also evident in the development of the balanced, well-integrated personality. I would argue that institutional controls, manifest in schools, bureaucracies and, physically, in organizational systems like land-use planning, reinforce this tendency. If this is the case, it is not surprising that small features of the urban landscape, like children's playgrounds and parks, and many domestic interiors, are also supervised, controlled spaces which signal exclusion. Because many of these controls are taken for granted or register negatively only in the world-views of others, like children, who have little power to influence the design of the spaces which they have to negotiate, we, that is, the dominant majority, are implicated in the perpetuation of the carceral control system.

As a metaphor, the carceral city, in which all of us are trapped, either as agents of domination or as victims, or both, has considerable value. The recognition that socio-spatial control is not restricted to particular and obvious sites of exclusion gives geographies of the asylum movement or the geography of prisons particular potency. Characteristic geometries and patterns of domination appear widely. Although there is not much room for agency in Foucault's thesis, since we are all apparently trapped in the carceral net, there is a connection between his structuralism and Freudian theories of the development of the self. Thus, Foucault's view that 'Men and women are always social creations, the products of codes and disciplines',[28] can be reconciled with the view of Erikson, Klein, Sennett and others on the production of the social self, where the other assumes both material and social forms which are articulated in rules of exclusion. This gives the thesis a more general significance, at least within developed, capitalist societies, although we should be wary of ignoring cultural difference and generalizing too far.

Foucault's analysis of social control is depressing. We are left feeling helpless. A similar conclusion might be reached from reading Stanley Cohen's *Visions of Social Control*.[29] Cohen challenges the view that exclusion, separation and isolation are necessary features of social control. He suggests, rather, that programmes designed to bring the 'deviant' back into the community result either in the reconstruction of group conflict at a different scale or more insidious modes of *inclusionary* control, which are less likely to be challenged because they are relatively benign and liberal. He maintains that

> when . . . boundary blurring, integration and community control take place, the result is that more people get involved in the 'control problem' . . . more rather than less attention has to be given to the deviance question. In order to include rather than exclude, a set of judgements has to be made which 'normalizes' intervention in a greater range of human life.[30]

Thus, more 'humane' penalties, like electronic tagging or community service, involve more people in the corrective and caring professions, they may involve the vetting of the families of recidivists, and they reduce awareness of control and the criminalizing of behaviour. 'Modern inclusionary social control becomes a system of "bleepers, screens and trackers", part of the "invisibly controlling city"'.[31] At least, the strong boundary between the prison as a site of exclusion and 'normal' space may serve to keep carceral punishment at a high level of consciousness, although diverting attention from exclusionary practices elsewhere in society.

The other strand in Cohen's argument is that when there is decarceration, the community replicates the territorial divisions that occur when there is a clear policy of separation for the mentally ill, mentally disabled or criminal. Thus, while asylums removed the mentally ill from the rest of the urban population, de-institutionalization isolates them also, particularly within inner-city areas. We have the creation of new ghettos, described in the North American context by Wolch and Dear.[32] Rather than being the inevitable geographical expression of de-institutionalization, however, it could be argued that this pattern reflects the inadequacy of community care. A properly funded programme of half-way houses, therapeutic treatment, employment provision, and so on could counter the tendency towards isolation and enclosure.

I think that we have to accept Cohen's argument that exclusion is not a necessary feature of social control. Exclusion is symbolically rich and it has provided an attractive theme for literature, both fictional and academic. The oppositions of inside/outside, pure/defiled, and strong spatial divisions are appealing and they do apply to some cases of socio-spatial control, but we have to recognize that social control can assume diffuse forms and may not be expressed in such stark terms geographically.

CONCLUSION

In this chapter, I have suggested that both space and society are implicated in the construction of the boundaries of the self but that the self is also projected onto society and onto space. Self and other, and the spaces they create and are alienated from, are defined through projection and introjection. Thus, the built environment assumes symbolic importance, reinforcing a desire for order and conformity if the environment itself is ordered and purified; in this way, space is implicated in the construction of deviancy. Pure spaces expose difference and facilitate the policing of boundaries. The problem is not solely one of control from above whereby agents of an oppressive state set up socio-spatial control systems in order to remove those perceived to be deviant and to induce conformity. A reading of Klein, Kristeva and Sennett suggests that exclusionary tendencies develop in the individual and that the exclusionary practices of the institutions of the capitalist state are supported by individual preferences for purity and order. Feelings of abjection are reflected in consumer advertising, for example. A rejection of difference is embedded in the social system.

One difficulty with this argument is that, despite the apparently universal nature of these processes, some people and some localities are more tolerant than others. In typologies of personality, the 'authoritarian' or 'foreclosed' is recognized as an exceptional category, and similarly, within the city, as Wolch and Dear indicate, there are contours of tolerance. Although there is no simple contrast between heterogeneous, accepting inner cities and homogeneous, rejecting suburbs, it could be the case that the *experience* of difference and mixing in social and spatial terms contributes to variations in the response to difference. Individuals are socialized into a variety of environments, both in the home and in the neighbourhood, and the forces of purification are not going to be equally effective in moulding all individuals, groups and localities. Furthermore, if we accept that people are active agents who think reflexively, there is always the possibility of springing the trap. Even the suburban couple in Sidcup whose home life is described in graphic detail by Cohen and Taylor may mock the bourgeois pretensions of their neighbours and create an enclave in which they are able to live a non-conforming life:

> The uniformity and predictability of it all might seem to induce an unshakeable sense of routine, a soul-destroying impression of the unmalleability of paramount reality. But when the door is shut against the night, and the two children are safely in bed, husband and wife turn to each other and laugh. They are subscribers to the new self-consciousness, apostles of awareness. Cynically, they deride those who share bourgeois arrangements with them, but who do not see the joke. Looking around the room they declare their awareness of their apparent suburbanity, and then with a delicious sense of their own distinctive identities, record their distance from such artifacts.[33]

Admittedly, the chances of this happening are not great, given the residential selection process and the fact that social and environmental homogeneity are mutually reinforcing, but the temptation to construct yet more social and spatial stereotypes should be resisted.

NOTES

1. Michel Foucault, 'Of other spaces', *Diacritics*, 16 (1), 1986, 22–27.

2. Edward Soja, *Post-modern Geographies: The reassertion of space in critical theory*, Verso, London, 1989, p. 119.

3. Allan Pred, 'The social becomes the spatial, the spatial becomes the social: enclosure, social change and the becoming of places in Skane', in Derek Gregory and John Urry (eds), *Social Relations and Spatial Structures*, Macmillan, London, 1985, pp. 337–365.

4. Soja, op. cit., chapter 6.

5. Anthony Giddens, *The Constitution of Society*, Polity Press, Cambridge, 1984, chapter 1.

6. Nicky Gregson, 'Structuration theory: some thoughts on the possibilities for empirical research', *Environment and Planning D: Society and Space*, 5, 1987, 73–91.

7. Giddens, op. cit., p. 368.

8. Steve Pile, 'Human agency and human geography re-visited: a critique of new models of the self', *Transactions, Institute of British Geographers*, NS, 18, 1993, 122–139.

9. A. Moos and M. Dear, 'Structuration theory in urban analysis, 1: theoretical exegesis', *Environment and Planning A*, 18, 1986, 231–252.

10. Michael Ignatieff, *A Just Measure of Pain: The penitentiary in the Industrial Revolution, 1750–1850*, Macmillan, Basingstoke, 1978.

11. Anthony Giddens, *Profiles and Critiques in Social Theory*, University of California Press, Berkeley, 1982, pp. 197–198.

12. Foucault, 1980, op. cit.

13. Michel de Certeau, 'Practices of space', in M. Blonsky (ed.), *On Signs*, Basil Blackwell, Oxford, 1985, pp. 122–145.

14. Fred Hirsch, *Social Limits to Growth*, Routledge, London, 1978, p. 82.

15. Julia Kristeva, in *Powers of Horror*, pp. 65–67, describes Mary Douglas's work on pollution/exclusion as fundamental. Mary Douglas, in *Natural Symbols*, devotes a chapter to Basil Bernstein. Bernstein, in *Class, Codes and Control*, vol. 1, acknowledges Douglas's influence. Both Douglas and Bernstein were clearly influenced by Emile Durkheim's thinking on the sacred and profane.

16. P. Atkinson, *Language, Structure and Reproduction: An introduction to the sociology of Basil Bernstein*, Methuen, Andover, 1985, p. 27.

17. Basil Bernstein, 'Open schools, open society?', *New Society*, 14 September 1967, 351–353.

18. Basil Bernstein, 'On the classification and framing of educational knowledge', British Sociological Association Annual Conference on Sociology of Education (reprinted in *Class, Codes and Control*, vol. 1, Paladin, St Albans, 1971, pp. 202–230).

19. I return to Bernstein and the idea of 'dangerous knowledge' in my account of excluded geographies in Part II of the book.

20. Andy Alaszewski, *Institutional Care and the Mentally Handicapped: The mental handicap hospital*, Croom Helm, London, 1986. Alaszewski trained as a geographer and then as a social anthropologist. His application of Mary Douglas's ideas in an analysis of the organization of interior spaces has been largely unacknowledged in geography.

21. Nanette Davis and Bo Anderson, *Social Control: The production of deviance in the modern state*, Irvington, N.Y., 1983.

22. The value of Foucault's arguments in *Discipline and Punish* for the development of a more nuanced socio-spatial theory has been recognized by Chris Philo, initially in *The Same and Other: On geographies, madness and outsiders*, Loughborough University, Department of Geography, Occasional Paper 11, 1987; by John Lowman, in 'The geography of social control: clarifying some themes', in David Evans and David Herbert (eds), *The Geography of Crime*, Routledge, London, 1989, pp. 228–259; and by Soja, op. cit., 1989.

23. Michael Walzer, 'The politics of Michel Foucault', in David Hoy (ed.), *Foucault: A critical reader*, Basil Blackwell, Oxford, 1986, 51–86.

24. Lowman, op. cit., p. 237.

25. Walzer, op. cit., p. 60.

26. Thomas Markus, *Buildings and Power: Freedom and control in the origin of modern building types*, Routledge, London, 1993, p. 97.

27. I discussed the dual role of planned settlements for indigenous minorities and Gypsies in *Outsiders in Urban Societies*, Basil Blackwell, Oxford, 1981, particularly chapters 9 and 11.

28. Walzer, op. cit., p. 61.

29. Stanley Cohen, *Visions of Social Control*, Polity Press, Cambridge, 1985.

30. ibid., pp. 230–231.

31. ibid., p. 230.

32. Jennifer Wolch and Michael Dear, *Landscapes of Despair*, Polity Press, Cambridge, 1987.

33. Stanley Cohen and Laurie Taylor, *Escape Attempts*, Allan Lane, London, 1976, pp. 32–33.

6

SPACES OF EXCLUSION: HOME, LOCALITY, NATION

People feel possessive about spaces but with varying degrees of intensity. Feelings of belonging and ownership attach to national territory, for example, and this plays an important part in geopolitics, but an exclusive nationalism, one which denies the sharing of national space by diverse cultures, is far from universal. At the local level, an exclusive right to domestic space is a fairly common claim, one that may be recognized in law. The home is personal space or family space, one which others enter only by invitation. Individuals and groups also feel territorial about neighbourhoods, but whether they do or do not depends on the location and the social composition of the area. Difference, as I suggested in Chapter 5, is less likely to to be noticed, less likely to be a source of threat, in a weakly classified environment like Jane Jacobs's Manhattan than it is in a strongly classified, purified space. However, the home, the neighbourhood and the nation are all potential spaces of exclusion.

Apart from describing cases of exclusion at all these levels, I hope to demonstrate how, to some extent, one conditions the other – how the locality and nation invade the home, for instance, providing cues for behaviour in families as they relate to their domestic environment. Spaces are tied together by media messages, by local rules about the appropriate uses of suburban gardens, by the state's immigration policies, and so on. Some of these connections may be illuminated by moving between the home, the locality and the nation rather than treating each spatial configuration as a discrete problem. I am not suggesting that these different sites of exclusion are substitutable or can be mapped onto each other because there are some clear distinctions to be made, in terms of the sources of power and the agents involved in the manipulation of space. There are, however, similarities in imagery and, more importantly, the attitudes of individuals, moving as social beings between the home and the locality, contribute to the shaping of social space and, less directly, have some influence on geopolitical relationships.

Relations between states are affected by the support of individuals for policies which may, for example, involve the demonizing of foreign powers or political leaders. If a fear of others is generated at the local as well as at the international level, as it was in the United States at the height of the cold war during the 1950s, for example, support for the government may be galvanized. Then, propaganda suggested that the feared communist might be your next-door neighbour. An acute awareness of and distaste for difference, the theme which Arthur Miller explored in *The Crucible*, his play about the seventeenth-century witch craze in Salem, Massachusetts, fed into geopolitical discourse. Propaganda, it is often argued, consists of 'techniques of persuasion which fail to abide by established and accepted norms of accuracy and truth. They seek to manipulate relationships in order to persuade people about a particular claim to truth.'[1] However, this kind of comment fails to acknowledge that people may be receptive to propaganda because it fits their world-view, one which already accommodates stereotyped images of others who populate the locality as well as distant foreign countries. Thus, there may be a reciprocal conditioning of attitudes expressed at the local and the international levels. This is not the only reason for examining exclusionary tendencies in the home, the neighbourhood and the nation. There are important problems associated with each locale as well as unexamined connections between them.

HOMES

I have suggested a connection between feelings of abjection and the rejection of difference by individuals, and the material environment. A fear of difference is projected onto the objects and spaces comprising the home or locality which can be polluted by the presence of non-conforming people, activities or artefacts. Here, I examine the home as one context for the exercise of power. I will argue that, where the desire for a purified environment is not shared by all members of a household, the home becomes a place of conflict, and there is some evidence to suggest that this kind of conflict may lead to behaviour problems, particularly in adolescence. Surprisingly, in academic disciplines where a recognition of this problem might be expected, particularly environmental psychology, the conflictual aspects of the home are not widely appreciated. 'The house as haven' is a much more common theme than 'the house as a source of conflict'. According to Lee Rainwater,[2] in American culture

> There is . . . a long history of the development of the house as a place of safety from both nonhuman and human threats, a history which culminates in guaranteeing the house, a man's castle [*sic*], against unreasonable search and seizure. The house becomes the place of maximum exercise of individual autonomy, minimum conformity to the formal and complex rules of public demeanor. The house acquires a sacred character from its complex intertwining with the self and from the symbolic character it has as a representation of the family.

The family has a unity; it is not problematic.

In human geography, the home as a locus of power relations has been neglected almost entirely,[3] but this is more a problem of recognizing legitimate systems of interest than a failure to see the home environment as one affected by territorial disputes. In geography, interest in residential patterns wanes at the garden gate, as if the private province of the home, as distinct from the larger public spaces constituting residential areas, were beyond the scope of a subject concerned with maps of places. If we are to understand the constitution of social space, this neglect of private space does not make sense. Admittedly, the home as a social and spatial complex has been seen as a legitimate concern in studies of rural housetypes, in the landscape tradition, and a few geographers, notably Peter Williams,[4] have attempted to incorporate the home in broader theoretical studies of the production of residential space. These have been marginal enterprises, however. As Williams argues, 'within mainstream sociology and geography, considerable attention has been given to the division of residential space into stratified territories and, to a lesser extent, to the impact that these have had on local social relations. By extension, housing and homes are implicated in the process, but though the former has been developed substantially the latter remains neglected', and later (p. 251), 'Despite the actual or implied importance ascribed to the home, our comprehension of this "locale" is extremely limited and in some respects quite distorted.' This chapter pursues the theme of power and the control of space but, to put the problem in perspective, I will comment further on the more popular view of the home as a private sphere, distinct from the public world of work and social intercourse.

THE HOUSE AS HAVEN

I think it is fair to say that most studies of behaviour in domestic environments in the last twenty years have depicted the home in benign terms. It is a refuge, a source of comfort in a world otherwise replete with tension and conflict, and the only environment in which individuals can function as autonomous agents. Middle-class North America and Europe have been the focus of attention and there has been little recognition of the fact that conceptions of home vary between cultures, but even with the limited scope of studies in environmental psychology, a rather singular view has been projected. Korosec-Serfaty characterizes the (middle-class suburban) home as follows: 'each one creates . . . a clearly evidenced universe . . . it is only in this spontaneous [*sic*] architecture . . . that a few signs of authentic living may be gleaned'.[5] Opportunities for individual expression are emphasized in Korosec-Serfaty's own account of the sub-territories of the house. She suggests, for example, that the living room 'bespeaks the dweller' but in socially acceptable terms. It is part of 'the being's mode of anchoring in space'.[6] Dovey,[7] drawing on Appleton's refuge-prospect landscape thesis, similarly stresses the security or anchoring provided by the home, 'a place of certainty within doubt, a familiar place in a strange world, a sacred place in a profane world'. She also argues that, because the domestic interior is dialectically related to the outside, the contrast between the positive experience of home and the negative experience of the wider environment gives the meaning of home greater intensity and depth. To give a final example of this celebratory view of the home, Cooper paints a picture, with gratifying intensity, of a home which is fulfilling and reassuring:

> made up of histories and possibilities. So, the empty house is full of spaces for the imagination, of hopes and opportunities. There is a dreamlike quality in the momentary association of things in the process of change, the accidental relationships of light, space and clutter. Endless alternatives exist in walls almost without traces. The empty space slowly fills, a kind of order is imposed, disciplining, choosing, fixing. The wide view becomes a picture on the wall, a backdrop for the contents of the room – we look increasingly inward toward the detail. But while the limited possibilities of empty, pristine spaces are lost, the changes are the acquisition of a history, a mirror to life . . . The home is a space replete with pasts and memories.[8]

This may seem like idealism but several fairly weighty theoreticians have given support to this conception of the home. There is, for example, Engels's view that the alienation which is a consequence of the capitalist relations of production creates a need to separate the domestic sphere from the world of work. The failure of this attempt to secure the home from the disruptive forces of capitalism is anticipated, however, in the Marxist contention that the commodification of everything is inevitable. Use has also been made of a psychoanalytical argument – that the home serves as a boundary of the self, the home secures privacy, and so 'Privacy mechanisms define the limits and boundaries of the self. When the permeability of those boundaries is under the control of a person, a sense of individuality develops.'[9] This is similar to Gaston Bachelard's view of the home as a happy memory, recalled in dreams and 'giving access to the initial shell which shelters the being'.[10] This provides the basis for a 'happy phenomenology of the home'. The experience of the home is not necessarily bound to the physical shell of the house because it can be dreamt anywhere, but Bachelard's *Poetics of Space* does convey the image of the home as comforting and restorative, the intellectual equivalent of *Homes and Gardens*, a British leisure magazine, or a Laura Ashley home furnishing catalogue.

Clearly, many people do derive satisfaction from the fabric of the home and the artefacts it contains. Space and objects together provide aesthetic experiences, they evoke memories. As Rochberg-Halton expresses it,[11] they form a *gestalt*, a sense of an organized whole which is greater than the sum of its parts, for the people who live with them, a *gestalt* which secures a sense of individuality – 'the placing of objects in rooms shows how different rooms in the house reveal different conceptions of self'. This is too cosy. While the home can have these positive symbolic qualities, it also provides the context for violence, child abuse, depression and other forms of mental illness. What is missing from 'the house as haven' thesis is a recognition of the polar tensions surrounding the use of domestic space, tensions which become a part of the problem of domination within families. They derive from the ambiguity of boundaries which some people have difficulty in resolving. Oppositions like inside/outside, clean/dirty, tidy/untidy are essential features of the dwelling and its sub-territories, but they are not stable or fixed. Dirt, for example, 'belongs' in the garden, but it invades the house; a tidy room becomes untidy through having its contents disturbed or having things brought into it which do not fit a conception of tidiness. Thus, the fear of pollution can be a constant source of anxiety and pollution is a consequence of the actions of others. The threat posed by dirt and disorder, if it is attributed to one partner or to children, may contribute to tensions between members of families. These tensions will be particularly associated with the routine use of space in the home.

While some people are relaxed about domestic disorder or dirt, a concern with order in the home can become obsessive. This in itself is a personality problem, but it could lead to other problems if the person or persons concerned about order and boundary maintenance constrain the activities of other family members unduly. This usually involves one or both parents controlling the activities of children, particularly the spacing and timing of activities in the home. Some psychological research suggests that excessive control, expressed as a rigid family regime with strong boundary maintenance, may translate into problem behaviours, like substance abuse, which is essentially transgressive.[12] There is no conclusive evidence for this, but a focus on problems in families does underline the importance of power, power to exclude children from certain rooms in the house or to regulate the timing of their activities. Consider, for example, this recollection of a childhood in Scotland:

> My brother and I used to race to the bus stop for the 1 o'clock bus back home to travel the mile and a quarter to a too hot dinner, followed by a sprint down the drive to catch the 1.30 p.m. bus back to school. I remember on one of these racings to and from the bus stop falling onto a newly tarred and stoned road but, dead or alive, I had to get to school . . . Bells, of course, rang between periods (45 minutes or so) to end play time. I remember when my leg was broken by a girl falling onto it that I was horrified to hear the bell ringing between my wailing, that I could not get up and join the serried ranks of children waiting for the janitor to direct them into the building. . . My parents were most wonderfully organized in running the hotel. The clock ruled. Its discipline was sacrosanct.[13]

The writer related her anxiety about time to parental control. She goes on:

> The time I spent over homework caused a lot of trouble between my parents and myself. I realize now that the internalized whip made me anxious about learning and I was too insecure not to work . . . This rubbed off on my daughter as well. A cruel inheritance which I regret. Father would come back late from a freemason's meeting and stand by the electric light, demanding that I stop wasting his money and go to bed. I would not. And when he threatened to take me away from school, I attempted suicide, putting on all the gas rings and howling silently into my school books. I am sure they never forgot this. Our relationship was a very guarded one, though this was never talked about.

This kind of regulation of time has its counterpart in the regulation of space, with inevitable exclusionary consequences. This is suggested by the experience of 'David', recalling his teenage years in the late 1960s:

Dad had his special chair near the T.V. and radio in the living room and the rest of the family had their own recognized seating places. The other living room was a 'best room', reserved for visitors but not attractive in any case because it was always freezing. Mother organized the house. She liked everything neat and tidy. Feet were not allowed on chairs and neither food nor friends could be taken upstairs. David's father was a bit unapproachable . . . Saturday night was always bath night; bed times for the children were fixed and strictly adhered to: the parents always went to bed at 10.30.[14]

One way to think about the exercise of power in families is to make use of Basil Bernstein's distinction between *positional* and *personalizing* families.[15] In the positional family, power is vested in the positions themselves, 'father', for example, signifying authority, and the relationship between the dominant member(s) and others in the family is authoritarian. Control is maintained through the imposition of arbitrary rules and instructions are given without explanation, for example, in exchanges like: 'do this' . . . 'why?' . . . 'because I say so'. In Bernstein's scheme, there would be an association between the positional family and the strong classification and framing of space. In the domestic sphere, dominant individuals would be concerned with the maintenance of spatial boundaries, keeping children out of adult spaces, for example, and with the temporal regulation of children's activities. Keeping control means maintaining clear, unambiguous boundaries.

In personalizing families, by contrast, parents are less concerned with exercising control through the use of arbitrary rules and, in general, children will have greater involvement in decision-making in the family (compare Bernstein's 'open curriculum'). Thus, decisions about the use of space, the demarcation of activities, are negotiable and, because boundary maintenance is of less consequence as a means of control than it is in a positional family, conflicts over the location of activities is less of a problem. It follows that the organization of space and time in the home will be in the form of weak classification and framing. Control by exclusion is associated with the positional family; control through appeals to the collective interests of family members is characteristic of the personalizing family.

The positional family would be more likely to generate anxieties because of its concern with strong boundary maintenance and exclusion. However, studies of family dynamics suggest that the issue is more complex than this. In particular, Minuchin has found that problems identified in family therapy are associated either with excessive *detachment* or excessive involvement, what he terms *enmeshment*.[16] In a development of this idea, Olson, Sprenkle and Russell distinguished between normal interaction patterns, which could be *separated*, where family members do not communicate a lot but they are not isolated from each other, or *connected*, where people communicate well but they are not living in each other's laps, and the abnormal patterns of detachment and enmeshment.[17]

Detachment signifies a lack of communication and isolation, a condition which may be associated with strong classification because strong boundaries will minimize interaction. Enmeshment, by contrast, is more likely to be associated with weak classification. If people interact excessively, this could be seen as a failure to recognize boundaries. 'Ashley', for example,

> was allowed to take his girlfriend into the best room but every ten minutes, a head would pop through the door and father would ask if they were alright. Ashley's elder sister received a thick ear for talking back to her Dad after she discovered him following her to see what she was up to.[18]

A third possibility is that dominant individuals in an enmeshed family may attempt to impose strong boundaries but fail, as might happen, for example, if a family is living in overcrowded conditions.

Space enters this scheme both as an enabling and as a constraining medium. Suppose we have a positional family in which the dominant parent tries to exclude the children from areas which the parent categorizes as 'adult space'. If there is a separate living room which can be isolated and labelled as an adult space, boundary enforcement will obviously be facilitated and children may feel excluded. If there are enough rooms in the dwelling to accommodate children's activities, giving them some autonomy, however, the potential for conflict will be less than in the first case. With less competition for space, there may be less interest in boundary maintenance. Exclusion may not now be an acute problem, although it would still be possible for a parent to control the timing of children's activities. Conversely, if a family with positional tendencies lives in a home with very few sub-territories, at the extreme in one room, boundary enforcement will be difficult, but it may appear imperative. As Sebba and Churchman maintain,

> [separated spaces] can have a stabilizing and regulating role at individual, group and community levels. Where no such fixed and clear boundaries exist, the territory may not serve this stabilizing function and may be, in fact, a source of conflict and tension.[19]

Thus, space may be important in the sense that the layout of the dwelling determines the number of physical boundaries and interior spaces which can be regulated or defended. The density of occupation of the home is also important because of its effect on opportunities for privacy. Olson, Sprenkle and Russell recognize separation as one characteristic of normal family interactions, and, certainly, privacy appears to be an important feature of healthy families in several surveys of home environments. Rochberg-Halton,[20] for example, stresses the importance of autonomy and privacy for children, and enmeshment is less likely if members of the family are able to secure some personal space.

Disentangling families and domestic spaces is difficult. It is likely that in most families parents occasionally exhibit positional tendencies and exclude their children in order to give themselves some privacy. Conversely, parents may fail to appreciate a child's need for privacy and may occasionally be excessively involved, but these occasional exercises of parental power are probably a part of the normal pattern of family relationships.[21] Consistent exclusion and rigid boundary enforcement or persistent intrusion into the lives and living spaces of children may, however, contribute to behaviour problems in children and adolescents. These are not solely problems of parenting. The home which is projected as a highly ordered and unpolluted space by purveyors of home furnishings does not provide a sympathetic environment for children. Exclusionary tendencies are exacerbated by commercial representations of ideal homes which render children a polluting presence. Otherwise, a lack of space may make life very difficult for parents who wish to avoid getting enmeshed with their children, as, for example, when families have to live in the confined space of a hotel room.

If the home environment and the locality affect each other, which is likely to be the case in socially homogeneous neighbourhoods where there is a collective sense of appropriate rules governing the use of space outside the home, anxieties about the use of space inside the home are likely to spill over onto the front lawn and the street. The exclusion of children inside the home may be accompanied by the exclusion of a larger cast of 'others' outside the home. This happens in places like Belle Terre, New York, or Candlelight Hills, Houston, Texas. In the 1970s, the Supreme Court ruled that the community of Belle Terre 'had a legitimate interest in excluding households whose members were not linked by "blood, adoption or marriage" in order to preserve family values, youth values, the blessings of quiet seclusion and clean air', making it a 'sanctuary for people'[*sic*].[22] Under the restrictive covenants at Candlelight Hills,[23]

No clothes line shall be constructed or maintained on any lot within the sight of the street or any adjacent lot. No fence shall be constructed on any lot out of any material except brick, wood or wrought iron without permission of the Architectural Control Committee.

The consequences of such an exclusive and ordered view of the space outside the home are nicely summed up in an essay by Taylor and Brower.[24] They identify the benefits which can be derived from controlling territory immediately adjacent to the home, including the lawns of fenceless North American suburbs, sidewalks and alleys. With some enthusiasm, they suggest that 'noisy or rambunctious children can be shooed off the lawn', loiterers can be threatened with police action if they do not get off the steps, and so on. By reducing noise, unwanted intrusions and unregulated activity in these outdoor spaces, 'the sense of security, orderliness, and the quality of life inside the house is enhanced . . . Worry is reduced. To put it simply, life inside the home is better and less intruded upon.' The otherness of children and 'loiterers' is emphasized in a later passage. They ask rhetorically 'How does territorial functioning contribute to the functioning of the immediate society? First, and perhaps most obviously, attempts to exert territorial control are part of the deviation-countering and vetoing mechanisms of a block behavior setting [*sic*].' The deviant include 'rowdy teens', and vetoing includes 'reprimands where the offending person (e.g. a street bum [*sic*] sitting on the curb)' is asked to leave.

Taylor and Brower reveal the real purpose of this exclusionary activity when they assert, without the slightest trace of irony, that

markers such as beautification and upkeep 'send a message to other residents'. By (to use a middle-class, suburban example) keeping the house freshly painted and in good repair, the lawn and shrubs neatly trimmed and the flower beds brightly planted, one is telling one's neighbors 'I've invested in where I live, I like my neighborhood, and I can be counted upon to help out if there is any local emergency.'

A debate with Richard Sennett would probably not have been very productive.

Inside the home and in the immediate locality, social and spatial order may be obvious and enduring characteristics of the environment. For those who do not fit, either children whose conceptions of time and space are at variance with those of controlling adults or the homeless, nomadic or black in a homogeneously white, middle-class space, such environments may be inherently exclusionary. Exclusions are often episodic, however, and I want to go on to describe cases of local rejection where images of place appear to play an important role, determining, it might seem, whether a group is excluded, accepted or ignored. Analysis of these cases contributes further to the discussion of stereotypes in Chapter 2.

LOCALITY AND OTHERNESS

Object relations theorists have identified the simultaneous feelings of repulsion and desire which attach to stereotyped others. Stereotypes, however, often include elements of place so that discrepancy or acceptance depend on the degree to which a group stereotype matches the place in which it is located. A group can be in the 'wrong' place if the stereotype locates it elsewhere. Alternatively, the group stereotype and the place stereotype may be in harmony: the group fits and is not a source of anxiety or hostility. This is the most familiar territory of geographical analysis, explored in some detail in relation to mental illness and disability, the location of noxious facilities, racial difference, and so on. Here, I want to consider three cases which demonstrate the variable way in which place can be incorporated in stereotypical representations of minority groups. Two cases are historical, concerning peripheral and marginalized minorities in France and England, respectively. The third concerns representations of the English countryside which provide the context for current debates about the otherness of groups which the state is attempting to contain or exclude, particularly from rural areas. What I want to suggest is that questions of same–other relations at the local level do not invariably involve drawing strong negative stereotypes of the other. This obviously does happen in some instances, but there is a subtle interplay of group stereotype and place which needs to be teased out.

THE RAGPICKERS OF PARIS

A 1903 government report on rag collection estimated that there were between five and six thousand ragpickers (*chiffonniers*) working in Paris.[25] Their work and settlements were recorded in the photographs of Eugene Atget between 1910 and 1914, and these images can be compared with (other) stereotyped accounts of their presence in the city. Ragpickers dealt with residues and were themselves residual, socially and spatially. As Nesbit observes: 'the bourgeois had already rejected the essence of that labor, garbage being his instinctive expression of negation'.[26] So, a space for the ragpickers was repeatedly created beyond bourgeois space to remove the threat of contamination. The bourgeoisie (according to Nesbit)

could only look upon the ragpicker with some horror, which is expressed by progressively expelling the *chiffonniers* from the city, first from the area around the place Maubert, then from the rue Mouffetard, then from the thirteenth arrondissement out to the zones and the faubourgs, for reasons, it is said, of Hygiene.

Horror was not the only reaction, however. There was also a fascination with the culture of the ragpickers. Although they were excluded and restricted in their activities by city ordinances, they also created their own spaces in the shanty towns and asserted their independence from the rest of the urban population through their appropriation of space although they were totally dependent on the larger society for their livelihood. This independence and the apparent disorder of their environment were, by comparison with the regulated existence of the majority, a sign of freedom and a source of envy. Nesbit cites a commentary by Privat d'Anglemont on the Cité Doré, a nineteenth-century shanty town which also provided shelter for the ragpickers: 'it is the land of happiness, of dreams, of easy come, easy go, by chance set down in the heart of a despotic empire'.[27] Nesbit, observing that the apparent freedom of the *chiffonnier* was a constant feature of bourgeois discourse, suggests that it was the foreignness of the culture which was the source of fascination. The freedom was that of an exotic culture, like that of the North American Indian. The image of freedom drew on romantic stereotypes which, I would suggest, were acceptable because of the spatial separation of the ragpickers and the rest of urban society. Their presence did not violate the boundary of bourgeois culture and Atget's photographic record of the *zoniers*, a term referring mostly to the ragpickers who occupied the spaces adjacent to the gates in the nineteenth-century walls, could be read as a record of the exotic. The photographs probably reveal his own fascination with the disorder and difference of a nomadic population.

The representation of ragpickers in Paris during this period suggest that the strength of the good and bad stereotypes was contingent on place. Their deviant presence in bourgeois space was signalled by ordinances which regulated refuse collection and restricted the opportunities for scavenging,[28] controls which were introduced as a part of the general cleaning up and sanitizing of the capital, and by the community's relegation to the urban periphery. Viewed in their shanty towns from a distance, however, or in Atget's photographs they gave a vicarious pleasure. The *chiffonniers* represented a romantic disorder, a desired freedom. It is not always this simple. The interplay of good and bad cultural stereotypes and stereotypical views of place can affect responses to non-conforming minorities in other ways.

GYPSY STEREOTYPES

Place has provided an important context for stereotypes of English Gypsies and, conversely, Gypsies constitute an occasional ingredient of place stereotypes – a part of the cultural 'colour' of Granada or the Camargue, for example. Feelings of hostility and anxiety, envy and desire are bound up with conceptions of place.

In addition to the racist stereotype, there is an enduring image of Gypsies in northern Europe as a constituent part of the rural scene, and it is here that the good stereotype is invariably located. In George Morland's painting *The Benevolent Sportsman*, for example (Plate 6.1), a Gypsy family are portrayed almost as a part of nature. Their bender tent, which provides a rudimentary shelter, is barely distinguishable from the undergrowth and the social order is suggested by the positioning of the family below the bourgeois gentleman on the horse. In one sense, the Gypsy family are portrayed as uncivilized, a part of nature and beyond the margins of civil society, but their association with nature also creates a sense of the exotic. They are mysterious and romantic, harmonizing with nature in a way which members of civilized society cannot. Similarly, modern idealizations of Gypsy life, used commercially to promote products as exotic and therefore desirable, associate Gypsy culture with the countryside (Plate 6.2). Both the Gypsies and the countryside are seen through a mist of nostalgia.

If the benign stereotype has a rural setting, however, Gypsies in the city are likely to appear out of place and to be represented in negative and malign terms. There is nothing fixed or stable about these images and place associations because the designation of a proper place for Gypsies depends on whose interests are affected by their presence and where the antagonist or supporter is speaking from. It is interesting, then, to note how the picturesque and malign images were employed selectively to support arguments about the

Plate 6.1 The Benevolent Sportsman, George Morland, 1792 (courtesy Fitzwilliam Museum, University of Cambridge)

Plate 6.2 The romance of Romany Creams. The use of a benign rural Gypsy image in the marketing of food (photo: author)

place of Gypsies in the countryside around London, particularly in Surrey, early in the present century.

DEBATES ON GYPSIES IN THE BRITISH PARLIAMENT

In parliamentary debates in 1908 and 1911 on proposed legislation affecting moveable dwellings, which largely concerned Gypsies, both good and bad images were drawn on in the construction of socio-spatial categories which served particular interests. In the House of Lords debates on the Moveable Dwellings Bill, we can identify defenders of the countryside, who may have been rural landowners, denigrating rural Gypsies, and others who were able to retain an image of Gypsies as a a part of nature, a minority which had a place in rural areas and against which strong exclusionary measures were not warranted.

In the 1908 debate, the construction of Gypsies as a polluting presence in the countryside, specifically in Surrey, was quite explicit. For example, Lord Farrer, a Surrey resident, claimed that he could remember when the commons of Surrey [were] entirely free from vans and Gypsies. One tenth of the county is common land and it is of enormous importance to keep the lungs of London free. Whereas it was formerly possible to enjoy those commons, they are now infested with tramps and Gypsies of every description.

He went on to justify his hostility towards Gypsies, asserting that 'the Gypsy living in harmony with nature' had almost disappeared, to be replaced by people who were deviant and threatening: 'the old-fashioned Gypsies are rapidly dying out and the commons of Surrey are now infested with tramps and nomads to such an extent that a real and serious evil exists'.[29] The Archbishop of Canterbury added some ecclesiastical weight to this argument and a contribution from Lord Belper had a similar moral tone. Both urged greater control of Gypsies to deal with the evil of increasing numbers and associated problems of education, sanitation and morality.

Such comments suggest that Gypsies were seen as a minority which did not fit in rural areas and their lack of fit made them polluting and threatening, as Mary Douglas might have argued. Others took a more sanguine view, however. The Earl of Crewe, for example, saw nothing threatening in rural Gypsies:

The real difficulty of the subject is . . . that there is a considerable number of people who would like practically to prohibit the existence of this nomadic population altogether. We often pass these people on the roads and we look at them with interest, not always, I am bound to say, without a certain degree of envy.

He was able to bracket them with the urban working class, thus de-emphasizing their difference (and deviance):

I think it is only fair to compare them, not with people who live in comfortable houses under good sanitary conditions but with those who live in the slums of the towns – people of the same class and who lead the same kind of existence which these people would lead if they were confined to urban areas.[30]

At the second reading of the bill in the House of Lords in 1911, the good and bad stereotypes were again used in arguments advocating different degrees of control. The moderate position was reflected in comments by Lord Clifford of Chudleigh, who saw that 'in the highly organized state of society in which we live . . . the free and romantic life of the Gypsy must be sacrificed somewhat for the well-being of the crowded community in which we live'.[31] By contrast, Lord Farrer, the most hostile speaker in the 1908 debate, still saw the Gypsy as an abomination: 'As a resident in the Home Counties, I must say that the evil has reached a pitch which really demands that some action should be taken on the part of the administrative authorities.'[32]

In these two debates, contrasting images of Gypsies in rural areas were presented in order to make a case for different degrees of spatial control. Those who denied Gypsies a place in rural society, who rejected the benign stereotype of Gypsies as a part of nature and substituted one of Gypsies as a criminal minority, advocated strong exclusionary pressures. To support this position, they coupled the language of pest control – reference to 'infestations' – and a Manichean opposition of good and evil. Lord Clifford of Chudleigh recognized that an Act informed by such views would have the effect of 'driving dwellers in moveable dwellings into counties where restrictions did not exist and, when those counties had in turn made similar protective provisions, of driving them, as the expression went, "off the face of the earth"'.[33] Those in favour of lesser controls used terms like 'wild', 'free' and 'romantic', which gave Gypsies a place in an imagined countryside, while some conceded that, in counties like Surrey, suburbanization was making this Arcadian conception of the countryside rather difficult to sustain.

OTHERS IN THE MODERN ENGLISH COUNTRYSIDE

The arguments associated with the Moveable Dwellings Bill are being replayed in the British parliament and elsewhere following the publication of the Criminal Justice and Public Order Bill in December 1993.[34] The public order clauses in this bill identify various others which the Tory government intends to confine and exclude through the use of penal measures. It is implicit in the wording of the bill that these others are primarily groups who appear discrepant in a purified English (and Welsh) countryside. John Major attacked New Age Travellers in his speech to the Tory Party Conference in 1992, reinforcing the bad stereotype which his predecessor had used in an earlier attack. His pledge to rid the country of New Age Travellers – 'New Age Travellers. Not in this age. Not in any age' (thunderous applause) – probably inspired some of the legislation, but the legal net is now cast much more widely to cover groups which move into the countryside and transgress by doing so. These other groups include Gypsies and Irish and Scottish Travellers who are not already accommodated on official sites,[35] ravers, hunt saboteurs and environmental protesters.

It is clearly the case that the criminalization of these disparate groups, which will become liable to heavy fines and imprisonment and whose movements will be severely curtailed by strengthening police powers, depends on the acceptance of negative stereotypes. The power of these stereotypes, however, relates to the context of their supposed transgressions, that is, a purified countryside. The countryside, it seems, belongs to the middle class, to landowners, and to people who engage in blood sports. This point was made in the House of Commons debate at the second reading of the bill. One Conservative supporter of the legislation, Sir Cranley Onslow, argued that :

> Part V [the section of the bill dealing with public order offences] and its provisions to strengthen the position of those people who want law and order to prevail in the countryside are an important departure from precedent. The creation of a new offence of aggravated trespass is a significant step forward that will be widely welcomed in all parts of the country *where people have become all too used to disorder, intimidation and violence prevailing and interrupting the lawful pursuits of those who live in the country, value it and want to continue with their countryside sports* [my italics].[36]

His vision of a peaceful rural community clearly excludes others who live in the English countryside but who do not participate in country sports, including many commuters, working-class families and those who value an alternative life-style, like New Age Travellers. This is symbolic politics which requires a singular and homogenized view of English rural society to sustain it. In turn, the vision of the rural which underpins this legislation is informed by a xenophobic nationalism – the English countryside as it is represented by politicians like Onslow, following a tradition which plays on images of thatched cottages and red-coated huntsmen, stands for England. This vision is by definition exclusionary.

This case suggests how a group, like New Age Travellers, can be denied a place in society through a particular construction of place. A rigid stereotype of place, the English countryside, throws up discrepant others. The presence of these discrepant groups reflects the anxieties of those in dominant positions who see their material interests threatened. Environmental protesters threaten the road programme on which major construction companies that support the Conservative Party depend, the migrations of New Age Travellers threaten rural landowners, and so on. The strengthening of the laws of trespass is an obvious means of protecting these interests. At the root of the problem is the assumption that the countryside belongs to the privileged. This assumption is challenged by the transgressions of the folk-devils targeted in this legislation. These groups are other, they are folk-devils, and they transgress only because the countryside is defined as a stereotyped pure space which cannot accommodate difference.[37]

NATIONAL IDENTITIES AND ALIEN OTHERS

The dominant image of the English countryside which underscored debates on the Criminal Justice and Public Order Bill touches on questions of national identity. The countryside, as it is represented by those who have a privileged place within it, is the essence of Englishness, so those who are excluded from this purified space are also, in a sense, un-English. It is those parts of national terrritory that are pictured as stable, culturally homogeneous, historically unchanging which are taken to represent the nation in nationalistic discourse. These are generally rural areas which stand in opposition to cosmopolitan cities. If they are cities, they are cities in the past, populated in the English case by pearly kings and queens, barrow boys and other honest working folk.

The cultural heterogeneity of the countryside or the city has to be denied in these fictional characterizations if they are to symbolize an imagined national community. This makes a black presence in the English countryside particularly potent because black people as 'immigrants' would be denied a place in a quintessentially English environment. This is a view that has been effectively challenged by the black British woman photographer Ingrid Pollard,[38] who has placed herself in cherished English landscapes in order to emphasize her exclusion. Paul Gilroy has similarly argued in regard to cultural racism that 'the family supplies the building blocks from which the national community is constructed',[39] but this family is the family of racist discourse. It excludes 'alien others'. Gilroy suggests that this racist denial of difference puts black women 'directly in the firing line . . . because they are seen as playing a key role in reproducing the alien culture and, secondly, because their fertility is identified as excessive and therefore threatening'. I think that we can recognize a number of building blocks or key sites of nationalistic sentiment, including the family, the suburb and the countryside, all of which implicitly exclude black people, gays and nomadic minorities from the nation. There is, in addition, a particular history which is invoked in times of national crisis which is equally exclusionary. In regard to British racism, Gilroy maintains that

> It has become commonplace to observe that the precious yet precarious Churchillian, stiff upper lip culture which only materializes in the midst of national adversity – underneath the arches, down in the air raid shelters where Britannia enjoyed her finest hours – is something from which blacks are excluded.[40]

I would argue, however, that it is not only blacks who are denied a role in heroic national episodes.

Occasionally, the national boundary may be breached by alien others, leading to panic. One example, which is worth noting because it was so trivial yet such a threat to national security, concerned a small group of Lovara Gypsies (referred to in the press and in parliament as 'German Gypsies') who arrived in Scotland in 1906 and spent the summer of that year travelling in England. This migration became a political issue of national importance for a few months, with one member of parliament talking about 'repelling their invasion and harrying them onwards' and another questioning whether they qualified for 'Christian hospitality'.[41] Their numbers were exaggerated – a plausible forty-eight became three hundred in one parliamentary debate. This particular panic has to be seen in the context of the 1905 Aliens Act which was designed primarily to restrict Jewish immigration from eastern Europe, but it had the effect of raising consciousness of foreigners in general and it fostered British fascism. German Gypsies, an alien speck in purified national territory, were an ideal target. What must have been experienced as a fleeting presence in some localities acted as a catalyst for the expression of nationalistic and xenophobic sentiments. Just as there may be a compelling urge to expel spies and witches from small, tightly knit communities, so the state in a modern society may be able to stir up feelings against an alien other, which represents no real threat to material interests or social stability, in order to sustain geopolitical relationships.

This kind of xenophobia, based as it is on a purified national identity, sits uneasily with the flows and cultural fusions which are generated by global capitalism. Cheap labour from the periphery has played a vital role in the creation of wealth in the core countries. The South has for a long time had a presence in the North. In practice, however, the contradiction between a racist nationalism and the imperatives of capitalist economies is denied. This is nicely illustrated in Michael Kearney's essay on the United States–Mexican border,[42] where the gulf between North and South is particularly stark. Here, it appears to be the intention of the state that the economic performance of California, Texas, New York, Illinois and other destinations for would-be migrants should not be impaired by exclusionary immigration policies. Reflecting on the policing of the border, Kearney observes that:

> The night scopes are but one component in a sophisticated, high-tech surveillance program that also includes motion sensors, searchlights, television cameras, helicopters, spotter planes and patrols in various kinds of boats and ground vehicles, all co-ordinated by computers and radio communications. The annual budget for this sector of the Border Patrol is millions of dollars, but no money has been allocated in recent years to repair the fence.

The policy of the United States government is not to keep out illegal immigrants but to regulate the flow. At the same time, the state has to accommodate nationalistic sentiments, to 'dissolve the ethnicity of its huddled masses' through a rhetoric that recognizes only 'Americans'. This is the experience of other core states also. Immigration policy makers in Britain and Germany use the same text and 'Fortress Europe' likewise embodies the idea of a mythical sameness which denies a place for many who are already there, providing cheap labour for core industries. The myth of cultural homogeneity is needed to sustain the nation-state, to ensure support for domestic and foreign policies which are conducted on behalf of the nation. Relations between groupings of states are similarly informed by notions of purity and defilement, good and evil, in order to secure solidarity in the conduct of international relations. It is convenient to have an alien other hovering on the margins.

GEOPOLITICS AND PURIFIED IDENTITIES

Historically, the largest-scale expression of purified identities was the division of a large part of the globe during the cold war. Ronald Reagan's Manichean perception of the Soviet Union as an 'evil empire' was in the tradition of cold warriors who, according to William Pietz,[43] construed totalitarianism as 'an unprecedented, radically novel phenomenon. It could only be regarded as alien to the truly civilized heritage of the west, as not only a monstrous but an illegitimate birth.' More particularly, totalitarianism was associated with a western view of *Russian* culture as essentially Oriental – 'The basic argument is that totalitarianism is nothing other than traditional Oriental despotism, plus modern police technology.' Pietz, drawing on the writing of George Kennan, demonstrates that American cold war discourse depended largely on the myth of Orientalism, which was assumed to characterize Russian culture – the mask of civility hiding an innate barbarism and duplicity. German totalitarianism during the Nazi period could be explained as an aberration, a relapse into barbarism. For 'Western Man, taught as he has been to look for hope and solace in the dignity of the human spirit, [totalitarianism] is surely a pathological, abnormal state'.[44] Thus, the 'free world' was safely distanced from the Soviet Union by invoking an *essential* difference between the West, guided by humanist principles, and an Oriental other. Any totalitarian or barbaric episodes in the West had to be explained away in order to sustain this division of 'the world' into good and bad. For those in the West with an interest in continuing the cold war, this was a necessary purification of global space, one which required an 'other'.

This view is mirrored in debates associated with identity politics within Europe, debates about a singular identity, a form of self-stereotyping which distinguishes a nation or a group of nations from others.[45] This is a particular concern during periods of rapid political change and instability. That elusive spatial entity 'central Europe', for example, has been defined in one direction in relation to a Russian other, what Kundera referred to in its totalitarian condition as 'the radical negation of the modern west',[46] and in the other direction, rather more ambiguously, in relation to a west European other (more materialistic and decadent?). This spatial relationship is not stable, however. Post-communist Russians also claim a European identity, as they did when Russia's European-ness was asserted by contrasting their identity to the 'barbarous Turk' early in the twentieth century.

Cold war rhetoric provides one demonstration of the value of a psychoanalytical perspective in making sense of geopolitics. As Gearóid Ó Tuathail has argued,[47] geopolitics is currently conditioned by the invention of Third World 'others' – Castro, Quadaffi, Noriega, Hussein – who stand for a threatening form of cultural difference and against whom the West has engaged in military or economic conflict. According to O Tuathail,

> to explain how elites are able to manipulate identities and tap into a collective unconscious to whip up popular enthusiasm for techno-wars in the Third World, geographers need to engage the discourses of psychoanalysis, particularly post-Lacanian feminist psychoanalysis. (Perhaps 'we' will learn who 'we' are.)

The images sustaining the cold war or those which accompany Third World techno-wars may seem distant from the socio-spatial nexus of suburban families in Candlelight Hills or Potters Bar, but echoes of otherness travel backwards and forwards, reinforcing neighbourhoods, providing electoral support for restrictive immigration practices and legitimating foreign policy. Although there is no necessary correspondence between exclusionary practices at these very different spatial scales, the people who enthusiastically accepted the 'Argie' stereotypes of government and media rhetoric during the British war against Argentina in the Falklands/Malvinas would be unlikely to welcome racial minorities in their neighbourhood or to support a more humane immigration policy. Local encounters or the fear of encounters with an other are informed by images of alien other worlds and these local–global connections which need to be teased out. At all these levels, there is a problem of denial and a need to embrace difference. To quote Julia Kristeva:

> Living with the other, with the foreigner, confronts us with the possibility, or not, of being an other. It is not simply – humanistically – a matter of being able to accept the other but of being in his place, and this means to imagine and make oneself other for oneself.[48]

NOTES

1. John Pickles, 'Texts, hermeneutics and propaganda maps', in Trevor Barnes and James Duncan (eds), *Writing Worlds: Discourse, text and metaphor in the representation of landscape*, Routledge, London, 1992, pp. 193–230.
2. Lee Rainwater, 'Fear and the house-as-haven in the lower class', *Journal of the American Institute of Planners*, 32 (1), 1966, 23–31.
3. The question of power in the family in relation to the use of domestic space is discussed in David Sibley and Geoff Lowe, 'Domestic space, modes of control and problem behaviour', *Geografiska Annaler*, 74B, 3, 1992, 189–197.
4. Peter Williams, 'Social relations, residential segregation and the home', in Keith Hoggart and Eleonore Kofman (eds), *Politics, Geography and Social Stratification*, Croom Helm, London, 1986, pp. 247–273.
5. Perla Korosec-Serfaty, 'The home from attic to cellar', *Journal of Environmental Psychology*, 4 (4), 1984, 172–179.
6. ibid.
7. K. Dovey, 'Homes and homelessness', in Irwin Altman and Carol Werner (eds), *Home Environments*, Plenum Press, New York, 1985, pp. 33–61.
8. M. Cooper, 'Making changes', in T. Putnam and C. Newton (eds), *Household Choices*, Futures Publications, London, 1990, pp. 37–42.

9. Irwin Altman, *The Environment and Social Behavior*, Brooks Cole, Monterey, Calif., 1975, p. 50.

10. Gaston Bachelard, *La Poétique de l'espace*, Presses Universitaires de France, Paris, 1981.

11. Eugene Rochberg-Halton, 'Object relations, role models and the cultivation of the self', *Environment and Behavior*, 16 (3), 1984, 335–368.

12. This is suggested by S. Holmes and L. Robins, 'The influence of childhood disciplinary experience on the development of alcoholism and depression', *Journal of Child Psychology and Psychiatry*, 28 (3), 1987, 399–415. However, research by Geoff Lowe, David Foxcroft and David Sibley on adolescent drinking behaviour in relation to boundary enforcement in the home was inconclusive (Geoff Lowe, David Foxcroft and David Sibley, *Adolescent Drinking and Family Life*, Harwood Academic, Chur, Switzerland, 1993).

13. Mass Observation, summer 1988, *Time*, woman correspondent (Mass Observation Archive, University of Sussex).

14. From a series of interviews conducted by Ian Warner, reported in Lowe, Foxcroft and Sibley, op. cit.

15. Basil Bernstein's family typology is described in P. Atkinson, *Language, Structure and Reproduction: An introduction to the sociology of Basil Bernstein*, Methuen, Andover, 1985.

16. Salvador Minuchin, *Families and Family Therapy*, Tavistock, London, 1974.

17. D. Olson, D. Sprenkle and C. Russell, 'Circumplex model of families and family systems', *Family Process*, 18, 1979, 2–28.

18. Interview by Ian Warner, in Lowe, Foxcroft and Sibley, op. cit.

19. Rachel Sebba and Arza Churchman, 'Territories and territoriality in the home', *Environment and Behavior*, 15 (2), 1983, 191–210.

20. Rochberg-Halton, op. cit.

21. 'Normal' here refers to behaviour which I assume to be characteristic of most British families. Cross-cultural studies are needed to put the British case in perspective. My impression is that in southern Europe, for example, there is less concern with boundary maintenance and the separation of adults and children in the home than in Britain. Conversely, the separation of adult spheres and children's spheres may be more pronounced in Sweden.

22. Lawrence Sager, 'Insular majorities unabated: Warth v. Seddon and City of Eastlake v. Forest City Enterprises, Inc.', *Harvard Law Review*, 91 (7), 1978, 1373–1425.

23. Constance Perin, *Everything in its Place*, Princeton University Press, Princeton, 1977, pp. 232–240.

24. Ralph Taylor and Sidney Brower, 'Home and near-home territories', in Irwin Altman and Carol Werner (eds), *Home Environments*, Plenum Press, New York, 1985, pp. 183–211.

25. This section is based on a study of Eugene Atget's photographs of Paris – Molly Nesbit, *Atget's Seven Albums*, Yale University Press, New Haven, 1992.

26. ibid., p. 171.

27. Privat d'Anglemont, 'La Ville des chiffonniers', *Paris Anecdote*, 1854, 218.

28. Nesbit notes that a prefect named Poubelle decreed that rubbish bins could be placed on the street only fifteen minutes before the time of collection – not much time for sifting through the contents. In current French usage, the word for dustbin is *poubelle*.

29. *Parliamentary Debates, House of Lords*, vol. 187, 1908, cols 452–453.

30. ibid., col. 461.

31. *Parliamentary Debates, House of Lords*, vol. 7, 1911, col. 99.

32. ibid., col. 103.

33. ibid., col. 99.

34. Criminal Justice and Public Order Bill, HMSO, London, 16 December 1993.

35. The bill proposes to relieve local authorities in England and Wales from the obligation of providing sites for Gypsies and other Travellers. This requires the repeal of Part 2 of the 1968 Caravan Sites Act, which was intended to solve the problem of continuous harassment and evictions of Travellers.

36. *Parliamentary Debates, House of Commons*, 11 January 1994, col. 46.

37. That the offence to 'the rural community' was caused by transgressive groups was also made explicit by Cranley Onslow. He suggested that

> The [hunt] saboteur movement has its roots not in the countryside but in the towns. Anyone who has seen busloads of Millwall supporters [a London football team] brought in to disrupt a hunt knows exactly what I am talking about.
>
> (*Parliamentary Debates, House of Commons*, 11 January 1994, col. 46)

38. Pollard's work is discussed in Phil Kinsman, *Landscapes of National Non-Identity*, Working Paper 17, Department of Geography, University of Nottingham, 1993.

39. Paul Gilroy, 'The end of anti-racism', *New Community*, 17 (1), 1990, 71–83.

40. ibid., pp. 75–76.

41. *Parliamentary Debates, House of Commons*, vol. 156, 1906, col. 482.

42. Michael Kearney, 'Borders and boundaries of the state and self at the end of empire', *Journal of Historical Sociology*, 4 (1), 1991, 52–74.

43. William Pietz, 'The post-colonialism of cold war discourse', *Social Text*, 1988, 55–75.

44. George Kennan, 'The sources of Soviet power', *Foreign Affairs*, 25 (4), 1947 (cited by Pietz, op. cit.)

45. Iver Neumann, 'Russia as Central Europe's constituting other', *East European Politics and Societies*, spring 1994, 349–369.

46. Cited by Neumann, op. cit.

47. Gearóid Ó Tuathail, 'Critical geopolitics and development theory: intensifying the dialogue', *Transactions, Institute of British Geographers*, NS, 19, 1994, 228–238.

48. Julia Kristeva, *Strangers to Ourselves*, Columbia University Press, New York, 1991.

Part II

THE EXCLUSION OF GEOGRAPHIES

Paul Feyerabend claimed that there is no idea which is not capable of improving our knowledge.[1] However, some ideas count for more than others. Power is not equally distributed in the knowledge industry, and those practitioners who have more of it have the capacity to marginalize or exclude the work of dissenters. Ideas which challenge established fields of academic knowledge can be disturbing. Some workers within the academy who have invested considerable energy in a particular theoretical position may become anxious and defensive when faced with arguments which challenge assumptions that they have lived happily with for most of their careers. Rather than embracing the new, they may reject it.

Established practices in academic *disciplines* – a term suggesting that problems are taught and researched according to sets of rules – favour the cautious and the conservative. Referring again to Feyerabend's anarchist manifesto for science, we might argue that innovation is made difficult by the habit of looking for confirmation of existing theories. He suggests that

> [the] consistency condition which demands that new hypotheses agree with accepted theories is unreasonable because it preserves the older theory, and not the better theory. Hypotheses contradicting well-confirmed theories give us evidence that cannot be obtained in any other way. Proliferation of theories is beneficial for science while uniformity impairs its critical power. Uniformity also endangers the free development of the individual.

As I hope to demonstrate in the following chapters, the kind of theoretical pluralism which Feyerabend advocates has not been characteristic of social science either. Rather, knowledge which has gained legitimacy, which has become part of the currency of academic communities, has often maintained its status to the exclusion of conflicting ideas. In the history of particular subjects, on occasions it has been seen to be necessary to defend the boundaries against threatening ideas, to affirm the sanctity of an academic territory. I will be suggesting that it is not just any new ideas which threaten established knowledge, however. Some have a particular potency if they raise questions about central social values. It may appear that they have the power to destabilize or even overturn social hierarchies. If critical ideas come from the oppressed, for example, from women or black authors in certain contexts, or from minorities whose world-view is informed by their sexuality, they may be considered dangerous because they challenge white, heterosexual male domination of the western knowledge industry. If the marginalized claim the centre ground, or argue that there is no centre ground, there is at least an implicit threat to the authority of the guardians of established knowledge.

The danger of the fragmentation or de-centring of knowledge is met by ritual practices which confirm borders and emphasize the sanctity of that knowledge which has been legitimated by dominant groups in the academy, such as professional institutes. Methods of analysis, styles of communication and theoretical consistency may be used as qualifying conditions for the membership of knowledge clubs. Restricting entry by employing ostensibly academic criteria is a way in which elites retain their power. However, local boundary disputes in academia have also to be viewed in wider contexts. Patriarchy or racial oppression, for example, may be implicated in the exclusion of knowledge, but they are clearly political and social issues which are far reaching.

There are certain parallels between the exclusion of minorities, the 'imperfect people' who disturb the homogenized and purified topographies of mainstream social space, and the exclusion of ideas which are seen to constitute a challenge to established hierarchies of knowledge and, thus, to power structures in academia. In both cases, there is a distaste for mixing expressed in the virtues of pure spaces and pure knowledge. In both cases, it is power – over geographical space or over the territory marked out by groups within an academic discipline – which is under threat.

The promise of post-modernism is that the barricades will be lowered, allowing the juxtapositioning and fusion of cultures and ideas, the creation of new hybrid forms. Cultural heterogeneity will be paralleled by a polyvocal academy. To some extent, the social sciences now reflect these changes, giving more space to feminist writing, to black perspectives, and so on, but everyone does not have an equal voice. This will require a radical shift in power relations, which has yet to happen. In fact, some new and apparently quite radical movements come with their own dialects which render them inaccessible and undemocratic. The cases which I examine in the following chapters, although they cannot be used to substantiate these assertions because they pre-date the post-modern, have echoes in some recent debates. Arguments between feminists and post-modernists, or between Marxists and feminists, for example, raise questions of authority and power. My case studies may be instances of the exclusion of knowledge which no longer occur, but I am not sure about this. Academic post-modern writing, although celebrating diversity, does not seem to me to represent a radical break with the past. In order to get involved in the production of knowledge maybe it is still advisable to play the game according to some well-defined (and restrictive) rules which give some ideas legitimacy and render others inconsequential or even dangerous.

NOTE

1. Paul Feyerabend, *Against Method*, Verso, London, 1988, p. 5.

7

THE EXCLUSION OF KNOWLEDGE

> it must not be forgotten that men of science, too, are but human and that most of them either belong by descent to the possessing classes and are steeped in the prejudices of their class, or else are in the actual service of government. Not out of the universities therefore does anarchism come.
>
> (Peter Kropotkin)[1]

This chapter is a reflection on the production of knowledge which touches on the same themes of power relations, boundaries, and the dangers and rewards which attach to boundary crossings that I examined in the first part of the book. I try to show how the success or failure of ideas is affected by the contexts in which they are produced. I use success and failure to refer to the degree to which ideas become a part of the wider currency of knowledge within a particular academic territory. First, I will comment briefly on some conventional routes into this problem.

In accounting for change in the constitution of scientific knowledge, philosophers of science have tended either to argue from the standpoint of reason, like Karl Popper, comparing the logic of different explanatory systems, or to attempt to build a sociology of knowledge, generalizing from the behaviour of individuals or groups working in 'the scientific community'.[2] The social or political contexts of scientific research have received much less attention. There are some contextual references in Thomas Kuhn's history of scientific change,[3] but more particularly in the work of Paul Feyerabend,[4] who has both a broader and a deeper view of the problem than philosophers like Karl Popper. Feyerabend's views, however, do not seem to have informed the debates about the nature of knowledge which have taken place within individual social sciences. Such debates within human geography, for example, have been predominantly introspective, exemplified by Ron Johnston's contributions.[5] They have been concerned primarily with the subject's own explanatory procedures, and, while they have drawn on Kuhn, the contextual character of knowledge production has been given a rather thin treatment.

The *leitmotif* for much of the discussion of scientific change has been provided by Kuhn's concept of the paradigm, that is, an accepted problem solution or exemplar, which guides research within particular scientific communities.[6] In geography during the 1960s and early 1970s, when some practitioners tried to steer the subject into the scientific mainstream, there was considerable enthusiasm for the idea of a dominant paradigm. Thus, Haggett and Chorley looked favourably on economics as a subject steered by a single, stable paradigm.[7] It was 'the most successful and sophisticated of social sciences' and they suggested that 'the early debates over the nature of economics had been replaced by rather stable – but largely invisible – rules as to what problems and methods economic science should cultivate'. Geography, by contrast, was judged by these authors to be in a state of revolutionary change – spatial science provided new paradigms which required established tenets of faith to be forsaken. There was a radical break with the past in that areal differentiation was incompatible with a universalizing, model-based locational analysis. In this case, the argument about what constituted useful knowledge, what was progressive, was concerned entirely with the relative merits of different paradigms.

It is no doubt legitimate to argue about the existence of paradigms or whether or not a subject is in a stable or revolutionary phase, but such introspection suggests that science is detached from society, in the sense that science is unconstrained by politics, real economics and unscientific belief systems. So, Barnes argued that:

> As one passes back into the history of science, one finds it (or if preferred, its cultural antecedents) less and less differentiated from the general culture. Internal/external definitions become more difficult to deploy on the basis of actors' own definitions; boundaries become more and more nebulous.[8]

Modern scientists may feel insulated from society by their esoteric practices, but the autonomy which Barnes recognized is more apparent than real. Socialization into the scientific culture may blind scientists to the real influences exerted by state agencies, for example, the military,[9] and by private capital. An emphasis on 'reason' or the conventions which follow from the adoption of a paradigm also distracts from power relations within institutions, where research may be steered in certain directions by heads of departments or research committees. It is notable that at no point in his essay on scientific revolutions does Kuhn refer to power relations within and between communities. We might compare his position on scientific change with Mary Hesse's view of the social scientist, whose 'evaluation of the data in terms of a commitment to a social theory would be more like taking a political stand'[10] which suggests that power has a central role in the production of social scientific knowledge.

The issue of what, within the academy, counts as legitimate knowledge is a complex one. The output of scientists and social scientists is not *determined* by dominant paradigms, although it is influenced by them. At the same time, the state and capital do not dictate what research is done in academic institutions, but they do encourage some forms of enquiry, primarily through the selective funding of research. I would agree with Barnes that 'Simply because science is *culture* and grows and changes on the basis of its cultural resources and possibilities, it does not respond *simply* to material or social influences and stimuli.'[11] There is clear evidence, however, that the work of some researchers is marginalized and results are suppressed. The exercise of power, both by groups and individuals within disciplines and by outside agencies, can be an important influence on the production of knowledge. In fact, the relationship between power and knowledge is more intimate than this, as Foucault recognized. '"Truth" is linked in a circular relation with systems of power which produce and sustain it'.[12]

What I am working towards in this and subsequent chapters is an observation on the power/knowledge nexus as it is demonstrated in the suppression, neglect or dismissal of ideas on the organization of social space produced by marginalized others, particularly black and female academics. However, in an attempt to make sense of these instances of exclusion, I will first examine the contexts of knowledge production, focusing on social relations within the academy and relations between academic institutions and society, those relations which define the loci of power in particular settings.

KNOWLEDGE AND POWER

Robert Nozick asked:

Why do they [philosophers, but we could substitute academics in general] strive to force everything into that one fixed perimeter? Why not another perimeter or, more radically, why not leave things where they are? What does having everything within a perimeter *do* for us? Why do we want it so? (What does it shield us from?)[13]

This compartmentalizing of knowledge is a characteristic of academia associated with the growth of more specialisms. It secures monopolies and insulates the purveyors of knowledge from the threat of challenging ideas. Compartmentalized knowledge, kept within secure boundaries, gives power and authority to those who peddle it. It is not just rational argument that determines the rejection or acceptance of ideas: fear, envy and covetousness also play a part. Feyerabend,[14] an anarchist like Nozick, argues that

Unanimity of opinion may be fitting for a church, for the frightened or greedy victims of some (ancient or modern) myth or for the weak and willing followers of some tyrant. Variety of opinion is necessary for objective knowledge. As a method that encourages variety it is the only method that is compatible with a humanitarian outlook.

If the exercise of power has a role in promoting or suppressing knowledge, how is this power expressed? Power in academia is reflected in the existence of hierarchies, that is, in the hierarchical organization of the purveyors of knowledge and in a ranking of *knowledges*. Practitioners defer to a small number of 'higher' authorities, canonical figures whose ideas are widely accepted as fundamental and who provide a discipline with its paradigms. Other knowledge is considered 'lower' if expressed in different idioms, for example, the writing of journalists but also writing in unconventional academic styles. Beneath the knowledge contained in popular literature, like newspapers and magazines, lies 'folk knowledge', which constitutes data for social scientists rather than being recognized as a form of conceptual understanding which can be considered on the same level as academic constructions of the social world. Thus, if we juxtapose Hagerstrand's log-transformed map of migration distances in southern Sweden and a Beaver Indian's map of hunting territories (Figure 7.1), the former would at one time (in the 1960s and 1970s) have been widely regarded within geography as an authoritative general statement but the latter only as an interesting expression of minority culture, an ethnographic fact, rather than both being considered as equally legitimate statements about spatial organization. Certainly, social scientists have to privilege their own analyses, expressed in appropriate codes, in order to justify their position as interpreters of the social world. This ordering of knowledge, what Barry Smart has called 'a scientific hierarchization of knowledges'[15] has resulted in the *disqualification*

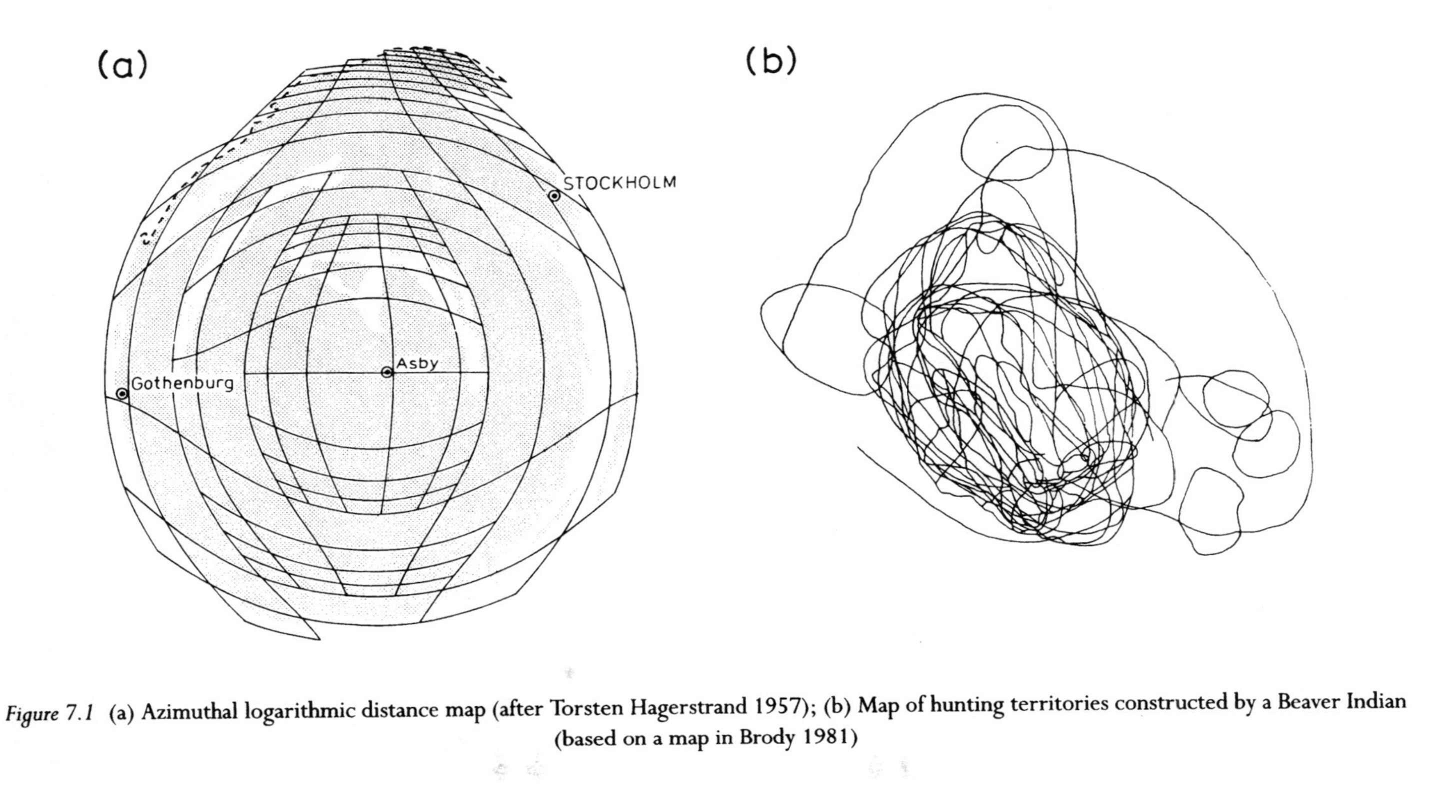

Figure 7.1 (a) Azimuthal logarithmic distance map (after Torsten Hagerstrand 1957); (b) Map of hunting territories constructed by a Beaver Indian (based on a map in Brody 1981)

of the low-ranking and popular forms and an *appreciation* of what Smart terms 'the politics of intellectual activity'.

Divisions of knowledge within hierarchical systems reflect increasing functional differentiation within developed economies. They have a history. Thus, in the early twentieth century, the boundaries between social sciences were blurred and the educational experience of prominent academics was often quite diverse. Robert Park, the Chicago sociologist, for example, studied engineering at the State University of Minnesota but, while there, he developed a strong interest in John Dewey's philosophy. He worked as a journalist after graduation, returned to academic life as a student of philosophy at Harvard, and submitted a Ph.D. thesis at Heidelberg University on 'The crowd and the public', a topic which would now be categorized as belonging to social psychology. Subsequent identification as a sociologist was for Park a necessary part of the political enterprise of gaining a place for sociology as a separate discipline at the University of Chicago. As I demonstrate in the next two chapters, this had important consequences for other social scientists who were not a part of Chicago sociology, those who did not qualify for membership of the community. Generally, as universities multiplied in order to service expanding industrial economies, academics had to stake a claim within them by appropriating areas of knowledge which had to be identified clearly by their content, methods and boundaries.

Recently, boundaries have tended to break down as practitioners have come to realize that many problems cannot be handled satisfactorily within established subject boundaries. Interdependencies have become more obvious, for example, in relation to environmental issues. Academic boundary crossing has been paralleled by greater political awareness of interdependencies in the social and physical world. As Geertz characterizes the current fusion of academic knowledge,

> rather than face an array of natural kinds, fixed types divided by sharp qualitative differences, we more and more see ourselves surrounded by a vast, almost continuous field of variously intended and diversely constructed works we can order practically, relationally, and as our purposes prompt us.[16]

This view is echoed by Spanos,[17] who sees an emerging de-centred field of knowledge where no single position is privileged and there is a free flow of ideas which dissolve former disciplinary boundaries. This is the end of logocentrism, the centring and disciplining associated with detachment and objectivity. Evidence of actual practice, however, suggests that the post-modern visions of Geertz and Spanos were rather optimistic. In a study of the behaviour of academics in a number of British universities, Becher observed that 'The deliberate freezing out of contributions which are seen [as] in some way threatening (usually because they purport to undermine an established ideology or school of thought) is not confined to any one knowledge area.'[18]

Bernstein's models of school curricula, outlined in Chapter 5, provide a good starting point for examining such aspects of the behaviour of academics, whose attitude to knowledge production may seem irrational to an outsider. Bernstein has expressed his core ideas in various ways, but the most general terms he uses to distinguish between those hierarchical and strongly bounded institutions or areas of knowledge, on the one hand, and non-hierarchical, weakly bounded institutions on the other, are strong and weak classification, the terms which I have used to describe the organization of social space. When knowledge is strongly classified:

> [It] is dangerous, it cannot be exchanged like money, it must be confined to special, well-chosen persons and even divorced from practical concerns. The forms of knowledge must always be well-insulated from each other: there must be no sparking across the forms with unpredictable outcomes. Specialization makes knowledge safe and protects the vital principles of social order.[19]

In this passage, Bernstein emphasizes the importance of the *control* of knowledge and, thus, by implication the perpetuation of the dominant value system – within a discipline, a sub-area or specialism, or an institution. Elsewhere, he notes the importance of 'differentiating rituals'. These rituals 'foster loyalty within such a segment and a degree of social distance from other equivalent segments. In this way . . . the units of a segmental, mechanical solidarity are created and re-created in symbolic forms.'[20] In this concern with boundaries, there is a similarity between academics and teenage gangs who tag their territory with graffiti and show their loyalty through the mixing of blood. Becher echoes Bernstein, arguing that:

> The more closely defined and better-defined the boundaries are between hard specialisms and the more tightly knit the groups associated with them, the easier it is to maintain the integrity of received doctrines by the ostracism or expulsion of internal dissidents and the refusal to provide entry permits to outsiders with dubious credentials.[21]

This could be Mary Douglas talking about witches and spies, but the comment is also particularly relevant to the accounts of black and female academics in the next two chapters.

To give two examples of this defensive, boundary-conscious attitude in modern geography, first, Minshull, clearly threatened by the injection of new ideas in geography, averred that:

Some of the changes in approach and purpose are so extreme, so diverse, that one would suggest that it is high time some new disciplines were hived off from geography before the work done by 'geographers' under the name of 'geography' becomes so diverse as to defy definitions.[22]

In a more open-minded paper on explanation in human geography, Johnston was still searching for 'what I consider [to be] truly geographical', and as Eyles and Lee remarked in their response to Johnston's paper,

[he] addresses geographical issues merely *as* geographical issues. Concepts, techniques, philosophies, reality are forced less into a 'structuralist mode' than into a geographical one. But the validity of such a transformation is ignored in the confidence and myopic vision of disciplinary imperialism.

The integrity of the subject, its distinctiveness and difference from other subjects, appeared to be Johnston's primary concern.[23]

In other fields, also, diversity, heterogeneity and disciplinary boundary crossing are regarded as undesirable. Becher suggests that appeals to epistemological unity are used to erect boundaries between the physical sciences. Thus:

one of my informants, a physical chemist, began by observing that the scientific background of physicists is based on deductive solutions, whereas that of the chemists is based on induction. This distinction, he suggested, pervades the practice and marks the common boundary of the two disciplines. Physics is limited in range: when a problem turns out to be too hard, the physicist marks it down as *dirty* and abandons it to the chemist. If the chemist also finds it too hard, he passes it on to the biologist, who adopts a phenomenological, rather than an inductive or deductive approach.[24]

This observation suggests a remarkable certainty about appropriate methods and the differentiating characteristics of scientific fields of knowledge. It is the antithesis of Feyerabend's prescription for scientific progress, namely, that 'Science is an essentially anarchic enterprise: theoretical anarchism is more humanitarian and more likely to encourage progress than its law-and-order alternatives.[25]

Because universities and other research institutions are so bound up with the state and supported directly and increasingly by capital, the tendency to promote and defend specialisms has to be seen partly as a response to the needs of these outside interests. It is also the case that there are more weakly classified areas of knowledge, blurred genres, where the need for a holistic approach which accommodates interdependencies is recognized. There is still plenty of evidence of boundary maintenance in academia, however, and this will not change while there is a commitment to maintaining knowledge hierarchies. The desire to maintain monopolies over areas of knowledge encourages ritual practices designed to protect the sacred status of established approaches to understanding.

LOST KNOWLEDGE, REJECTED KNOWLEDGE

Although the production of knowledge as a form of social practice implies differential power among practitioners, it is virtually impossible to demonstrate that ideas which challenge firmly established interpretations of the social and physical world have been deliberately suppressed. Academics within specialist areas may hold conversations only with each other but otherwise, as Becher suggests, 'talk past each other', a tendency which may now be encouraged with the proliferation of specialist journals. Ideas which, retrospectively, appear to challenge accepted thinking may simply not register. In order to secure a place in the knowledge business, it may be necessary for dissenting groups to launch their own journals and form their own study groups, as is evident in the recent past with Marxists and feminists in social science. This may compound the problem, however. If they do not become a part of mainstream discourses, they may remain invisible or, at least, be considered irrelevant and unimportant. The problem is analogous to one in spatial modelling. A model like Ernest Burgess's representation of the organization of social space in the western city attained the status of a universal statement with the effect that other readings of the city, representing other world-views, were not seen. Those who thought that Burgess's interpretation was valid did not necessarily have to consider and reject alternative interpretations because the idea of a multiplicity of equally valid world-views was alien to their universalizing, scientific perspective on the world.

In order to demonstrate connections between the production of knowledge, power relations and the wider context of knowledge, I will describe three

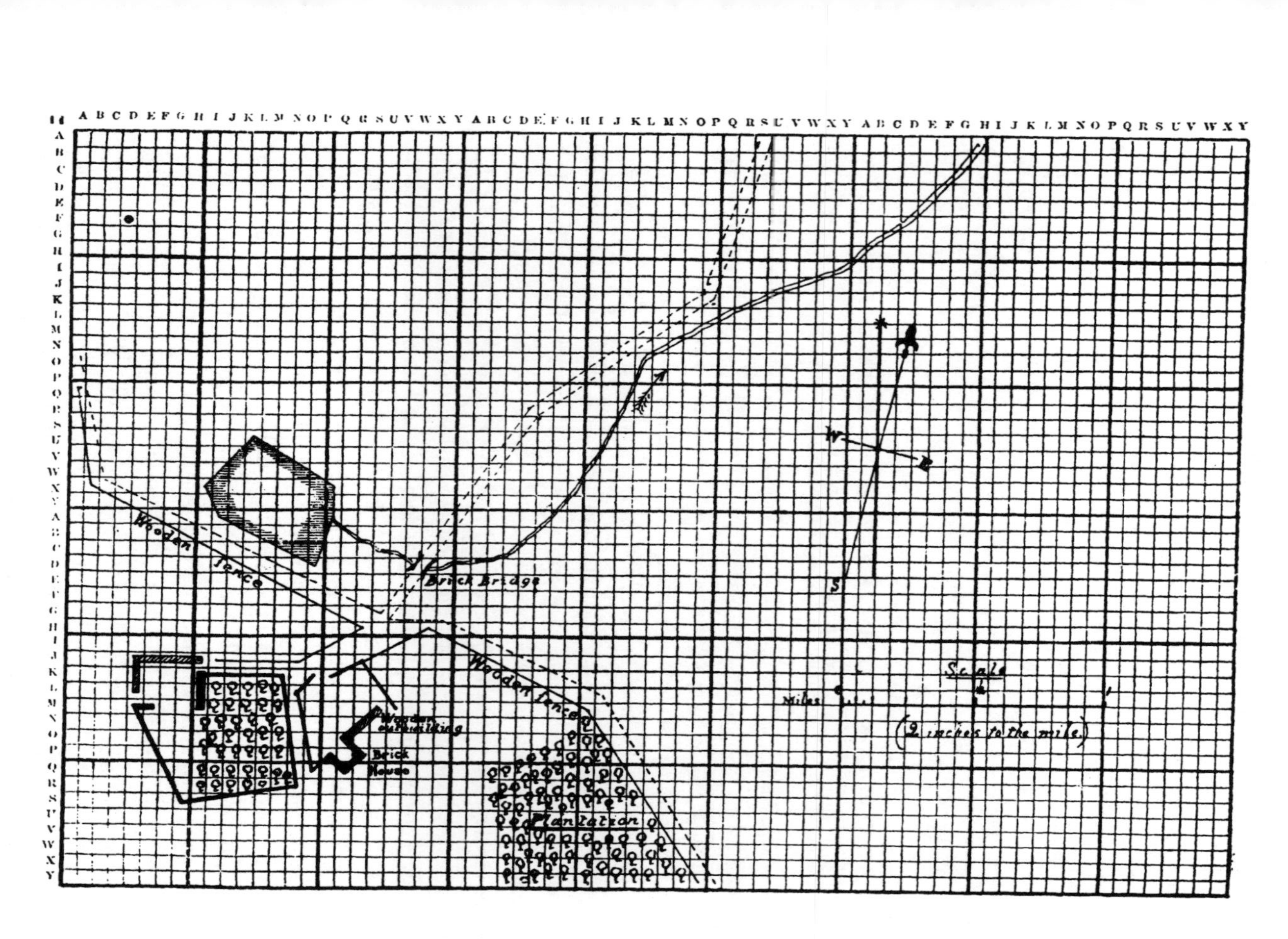

Figure 7.2 A map compiled from information transmitted by telegraph (source: Willink 1885)

Plate 7.1 A portrait compiled from information transmitted by the map-flapping method, using a grey-shading scale (source: Willink 1885)

instances of ideas which failed to gain any currency, but for different reasons. The first was plucked at random from the library shelves. It may be trivial, and it was unexceptional and probably useless, but it does provide a nice contrast with the second and third because it is most unlikely to have threatened existing power relations.

In an article entitled 'Map-flapping', which appeared in *Longmans Magazine* in 1885, H. G. Willink described a method for transmitting mapped data over distance.[26] It consisted of a code which included map coordinates, given by letters, and strings of letters which indicated features on the map. Thus, a part of the code might be:

Buildings, A; Brick, B. (This informs the receiver (1) that in the present message all lines or points, until a fresh descriptive word is signalled, are to represent brick buildings and (2) that in any subsequent part of the message the letter A, sent by itself, means 'Buildings' and the letter B means 'Brick'.)

Other letters in the code indicated that subsequent points were to be joined up, and so on, so all categories of object, their dimensions and location could be described by sequences of letters. The information, in the form of strings, was then signalled and the map reconstructed by a draughtsman at the receiving end (Figure 7.2). Letter-coding a shading system allowed more elaborate images to be transmitted, such as the portrait in Plate 7.1.

One practical problem with map-flapping was that it was very time-consuming, signalling a simple map taking maybe an hour, with a further forty minutes for complete reconstruction, even with the draughtsman at work while the message was coming in. The applications would have been primarily military, but the slowness of transmission probably rendered it impractical. The method had been promoted by army personnel for military use and, in principle, it bears a close similarity to modern computerized terrain mapping. It is doubtful that it represented a challenge to established thinking on reconnaissance. Map-flapping is hardly likely to have threatened existing practices and its failure to catch on is probably explained by its uselessness. The method, ingenuity apart, was a product of its militaristic times and the appearance of this article in a literary journal, although rather incongruous, would not have startled its readers unduly.[27]

The second case is quite different in that it illustrates the importance of threat in stifling new thinking. It concerns the the response of the medical, scientific and theological establishments to phrenology, the 'science of neuro-anatomy which was developed by two German physicians, Gall and Spurzheim, in the first two decades of the nineteenth century'.[28] Phrenology was a method of diagnosing personality traits from an examination of cranial contours. Cerebral function was inferred from the contours of the skull but the 'science' was more than a diagnostic tool. As Shapin points out, phrenology was anti-academic, a deviant development in neuro-anatomy from the medical establishment's perspective, and it was materialist, asserting the non-existence of an immaterial soul. The centre of the controversy was Edinburgh. Here, particularly, phrenology was viewed as heresy because Edinburgh was the major European centre of clinical medicine at the beginning of the nineteenth century. More generally, it was a prominent university city and a stronghold of the Presbyterian Kirk. All these interest groups – the medical establishment, the university and the church – were offended by the claims of phrenology.

The controversy surrounding phrenology illustrates the importance of power gained through the monopolization of knowledge and practical skills in academia and the professions and the danger to this monopoly represented by competing theories and methods. First, the attraction of phrenology was that it demystified some established practices and challenged the medical establishment. As Shapin observes:

> phrenology, like other recognized sciences, was based upon observation and, therefore, upon capacities common to *all* competent members of society. It was not, like the mental philosophy of the universities, based upon the mystifying and biased 'method' of introspection. Thus, phrenology could be seen as a commentary on the access of social groups to certifiable natural knowledge. It redefined the boundaries which separated disaffected social groups from a source of crucial legitimation. An observation-based psychology opened up a social system, by opening up access to resources for its criticism and change.[29]

In other words, phrenology made medical science more democratic. Although phrenology was discredited by objective, disinterested science, the initial attacks on it as a theoretical system were motivated, in Edinburgh at least, by the threat it posed to an established theology, through its denial of the immaterial soul, and to the surgeons and clinicians who denied the possibility of inferring cerebral function from cranial contours. The phrenological method conflicted with established practice, including dissection to identify organic function. For its proponents, outsider intellectuals, phrenology was attractive, according to Shapin, 'as a symbolic system which could be used as a juxtaposition to the institutionalized mental philosophy of local elites'. More generally, the reaction of the various Edinburgh establishments to the phrenologists illustrates Smart's argument about the hierarchical ordering of knowledges. Those forms of knowledge which are widely accessible have to be viewed negatively, decreed 'unscientific' by the academy in order to maintain the knowledge hierarchy and, thus, the power and privilege of those at the top. An ironic footnote to the phrenology debate is that the focus in Gall and Spurzheim's work on the cerebral localization of function anticipated knowledge of brain anatomy dating from the late nineteenth century. Retrospectively, it is evident that it had some scientific merit.

The third case brings us closer to the particular social issues which are the substance of the next chapter. Social knowledge can be more potent than physical or biological knowledge when it touches on visions of a moral order, although there are instances of physical theory, like Darwinian evolutionary theory or Eysenck and Jensen's theories of IQ inheritance, which have also proved highly disturbing in a moral sense. 'Dangerous' knowledge embodies values which call into question the moral basis of dominant models of society. Power is again a central issue because it is the establishment which has the power to define legitimate knowledge and to identify competing truth claims as deviant and dangerous.

This is illustrated by a controversy in Australian education, described by Johnston.[30] It concerned a secondary school teacher who was charged with improper conduct by the New South Wales Education Department following a complaint from a parent that he gave an 'unsuitable lesson'. There were, in fact, two allegations, first, that his teaching was inappropriate and second, that his relationship with girl pupils was too familiar, but the two are connected. On the first count, he allegedly read the class a story in which a strip-tease artist was sexually assaulted by a group of drunken young men. He then asked the girls in the class to put themselves in the place of the stripper and describe her feelings. In the story, the young men were middle-class old boys of a private school and the woman strip-tease artist was working class, in and out of jobs and doing this kind of work because she was desperate for the money. Johnston suggests that the story inverted the stereotyped view of class sexual behaviour. The violence of the middle-class men overturned the moral order, according to which violence and promiscuity are the prerogative of the working class.

The allegation about this lesson was coupled with accusations that, in English and history lessons, the teacher dwelt excessively on violence, horror and death. Johnston suggests that the charges were connected with allegations about familiarity with girl pupils in the sense that in his teaching he 'transmitted knowledge that was morally anomalous' and in that in his personalizing rather than positional approach to his pupils, the teacher 'stressed empathy with those caught up in a moral conflict and a critical, questioning approach to conventional moral categories'. He argues that the violent and sexually deviant, 'Girls committing suicide, strippers, hangmen, freaks, assassins, criminals . . . occupy the borderlands of our society.' Thus,

> To recognize their existence in the classroom and to adopt a strategy of appreciation and empathy means reducing the distance between the normal and the abnormal . . . Such an approach is likely to be very threatening to those whose moral universe is inhabited by firm, unambiguous rules and rulers.

In this case, then, the control of knowledge has an important moral dimension. Excluded knowledge is a narrative concerning groups and individuals who have been relegated to the margins of society because the values they represent undermine the moral consensus. This knowledge becomes dangerous and threatening when it is brought to the centre and presented as a legitimate perspective on social relations. It threatens to dislodge the centre, defined here by social knowledge which its advocates claim represents a moral consensus, and, thus, it becomes imperative to repel dissident thoughts or, in this instance, the messenger.

CONCLUSION

I have described three cases which vary considerably in their significance. The map-flapping exercise was probably inconsequential. It is unlikely to have represented a threat to established practice and was probably forgotten rather than suppressed, simply because it lacked utility. The history of ideas is no doubt littered with similar instances. The dispute over phrenology is more significant because it illustrates the tendency of dominant groups in the professions to exclude ideas which threaten their position. A single way, theoretically and practically, was presented as the professionally correct one and attempts were made to discredit an alternative view which challenged the monopoly of the establishment. The third example points to the importance of the control of knowledge by state institutions, particularly because social knowledge is construed as a moral order, a constructed truth given legitimation by the state. In order to protect officially sanctioned knowledge, the dissemination of ideas has to be policed. The educational system as characterized in Johnston's paper is a part of a panoptic policing activity, 'a supervision with the view to repressing, mastering, and domesticating (cultivating) the "errant" impulses that threaten to disrupt the authority and hegemony of the privileged majority discourse'.[31]

References to an 'establishment' in this discussion suggest a monolithic academic or other professional group repelling attacks by dissident thinkers. This may be an apposite description of some professions, like medicine in some historical periods, and of hierarchical and highly regulated school systems, like the French system and, increasingly, the British one following the introduction of a national curriculum. As Becher has suggested, however, universities in western societies are more loosely structured, with the content of subjects determined less from above than in secondary schools. There are controls, but they are exercised with greater subtlety by professional gatekeepers. The cases I describe in the next two chapters are ones where exclusionary forces appear to have been at work in academia. Admittedly, exclusionary processes are often difficult to document and disentangle – according to Hacking: 'Nobody knows this knowledge; no one wields this power.'[32] What I will suggest, however, is that when dissenting, radical ideas are produced by members of social groups who are themselves marginalized and excluded from centres of power, the threat to the establishment may be more tangible than when it comes from within. At least, the immediate sources of power, the immediate determinants of legitimate knowledge, may be knowable. The larger power structures in which particular academic conflicts are embedded can be more elusive.

NOTES

1. Peter Kropotkin, *Modern Science and Anarchism*, Simian, London, no date.

2. According to Barry Barnes, Thomas Kuhn's perspective on scientific investigation, 'so often described wholly in terms of the "reason" and perception of the isolated individual and his [*sic*] experience, is presented as a complex interaction between a research community, with its received culture, and its environment' (*T. S. Kuhn and Social Science*, Macmillan, London, 1982). This is tautologous, however. Kuhn's research communities constitute the environment of individual researchers (see the postcript to Kuhn's *The Structure of Scientific Revolutions*, 2nd edn, University of Chicago Press, Chicago, 1970). This restricted view of the context of scientific investigation probably reflects Kuhn's faith in the idea of scientific logic as the ultimate determinant of scientific change. As Merquior remarks: 'this Darwinian picture of paradigm struggle seems to harbour a residual homage to the objective, immanent logic of scientific argument' (José Merquior, *Foucault*, Fontana Press, London, 1985, p. 38).

3. Kuhn, op. cit., Chapter 10, 'Revolutions as changes of world view'.

4. Paul Feyerabend's two books on epistemology and the history of science, *Farewell to Reason*, Verso, London, 1987, and *Against Method*, Verso, London, 1988, are anarchistic critiques of the production of scientific knowledge in the West.

5. Particularly Ron Johnston, *Philosophy and Human Geography*, Edward Arnold, London, 1986.

6. Thus, Chalmers maintains that 'Normal scientists must presuppose that a paradigm provides the means for the solution of puzzles posed *within* it.' Barnes, 1982, op. cit., p. 46.

7. Peter Haggett and Richard Chorley, *Models in Geography*, Methuen, London, 1967, p. 27.

8. Barry Barnes, *Scientific Knowledge and Sociological Theory*, Routledge and Kegan Paul, London, 1974, p. 121.

9. For a good example of this, see Neil Smith, 'History and philosophy of geography: real wars, theory wars', *Progress in Human Geography*, 16 (2), 1992, 257–271.

10. Cited by David Hoy, 'Power, repression, progress: Foucault, Lukes and the Frankfurt School', in David Hoy (ed.), *Foucault: A critical reader*, Basil Blackwell, Oxford, 1986, p. 124.

11. Barnes, 1974, op. cit., p. 121.

12. Hoy, op. cit., p. 124.

13. Robert Nozick, *Anarchy, State and Utopia*, Basil Blackwell, Oxford, 1974, p. xiii.

14. Feyerabend, 1988, op. cit., p. 32.

15. Barry Smart, 'The politics of truth and the problem of hegemony', in Hoy, op. cit., p. 164. The presumed superiority of 'modern' objective thinking over 'traditional' modes of thought was examined by Lévi-Strauss in *La Pensée sauvage* (1966). 'Every civilization', he argued, 'tends to overestimate the objective nature of its own thought' (cited by Barnes, 1974, op. cit.).

16. Clifford Geertz, 'Blurred genres: the refiguration of social thought', *American Scholar*, 49, 1980, 166.

17. William Spanos, 'Boundary 2 and the polity of interest: humanism, the "centre elsewhere", and power', *Boundary 2*, 12 (3), 1984, 173–214.

18. Anthony Becher, *Academic Tribes and Territories*, Open University Press, Milton Keynes, 1989, p. 64.

19. Basil Bernstein, cited by P. Atkinson, *Language, Structure and Reproduction: An introduction to the sociology of Basil Bernstein*, Methuen, Andover, 1985, p. 28.

20. ibid., p. 30.

21. Becher, op. cit., p. 60.

22. Cited by Peter Taylor, 'An interpretation of the quantification debate in British geography', *Transactions, Institute of British Geographers*, NS, 1 (2), 1976, 129–142.

23. Ron Johnston, 'On the nature of explanation in human geography', *Transactions, Institute of British Geographers*, NS, 5 (4), 1980, 402–413, and a comment by John Eyles and Roger Lee, 'Human geography in explanation', *Transactions, Institute of British Geographers*, NS, 7 (1), 1982, 117–122. Johnston is still accused of being overly concerned with the boundaries of human geography. As Andrew Sayer comments:

> There is, unfortunately, more than a hint of disciplinary imperialism in the bid to get more space into social theory. Bourdieu's *Homo academicus* provides sociological explanations of this kind of phenomenon in relation to struggles within the status hierarchies of academia; it makes depressing reading for geographers.

He concludes, 'So: Down with all forms of disciplinary imperialism/parochialism, be they geographical, sociological or whatever! Long live postdisciplinary studies' (Andrew Sayer, 'Realism and space: a reply to Ron Johnston', *Political Geography*, 13 (2), 1994, 107–109).

24. Becher, op. cit., p. 39.

25. Paul Feyerabend, 1988, op. cit., p. 5.

26. H. Willink, 'Map-flapping', *Longmans Magazine*, 7, 1885, 404–415.

27. A comparison with geographical information systems (GIS) seems apposite. Whereas the military applications of map-flapping were quite explicit and unexceptional in a nineteenth-century imperial power, the military applications of GIS are largely undiscussed. As Neil Smith, op. cit., remarks, 'the academic advocacy of GIS seems deliriously detached'. The ethical issues associated with GIS would become more transparent if the enormous financial support of the military, particularly the United States Department of Defense, and its devastating effect in war were acknowledged.

28. Steven Shapin, 'The politics of observation: cerebral anatomy and social interests in the Edinburgh phrenology disputes', in Roy Wallis (ed.), *On the Margins of Science: The social construction of rejected knowledge*, Sociological Review Monograph 27, University of Keele, Keele, 1979, 139–178.

29. ibid., p. 146.

30. Ken Johnston, 'Dangerous knowledge: a case study in the social control of knowledge', *Australian and New Zealand Journal of Sociology*, 14 (2), 1978, 104–113.

31. Spanos, op. cit., p. 183.

32. This quotation is a comment by Hacking, op. cit., on Foucault's thesis on power/knowledge, developed in *The Archeology of Knowledge* and other writing. The gist of his argument is that power is exercised often unwittingly by many agents who constitute a web of control. Little acts and gestures contribute to the suppression of people (and ideas), but it is not usually a case of conscious suppression: 'those ruling classes don't know how they do it, nor could they do it without the other terms in the power relation – the functionaries, the governed, the

repressed, the exiled – each willingly or unwillingly doing their bit', as Hacking puts it. The exercise of power over ideas in academia may be more naked than Foucault suggests, although here, as elsewhere, exclusions or suppressions have to be understood in terms of many interconnected acts, some of which have unintended consequences.

8

W. E. B. DUBOIS: A BLACK PERSPECTIVE ON SOCIAL SPACE

He was not born a radical though he was born black.

(Jack Moore)[1]

The whiteness of human geography is a problem for a subject which concerns itself with the diverse experiences of the world's population. The whiteness of practitioners in the main centres of geographical research in Europe, North America and Australasia makes it difficult for 'other voices' to be heard in geographical conversations. This is manifest in the failure to acknowledge the contributions to an understanding of social space which black authors have made. Black world-views have come across to a white readership through the work of authors like James Baldwin and Franz Fanon, as well as social scientists like bell hooks and Paul Gilroy, but black fiction and social science writing have not informed much geographical writing.[2]

The post-modern enthusiasm for difference seems rather unconvincing when the presence of authors other than white ones in the literature is such a meagre one, but it is clearly difficult to shake off Eurocentric and imperialistic legacies in a subject whose history is so much bound up with colonialism. Research by white social scientists on the South still has a colonial tinge. I have heard researchers in south-east Asia, for example, using the possessive 'my' or 'his' when referring to peoples of the region, as if they were their academic property. Similarly, in the developed North, black inner-city communities are still unable to represent themselves, and their experience is interpreted primarily by white social scientists. In this chapter, I am concerned with the absence of a black voice in urban studies and I try to signal this by considering the neglect of the work of W. E. B. DuBois. He is a major figure in black American history whose work on the city was, I think, a significant contribution to the understanding of social space, one which was effectively eclipsed by Chicago sociology and which failed to find a way into the mainstream of urban geography.

W. E. B. DuBois is best known in the United States as one of the founders of the National Association for the Advancement of Colored People (NAACP), but his early interests were in history and sociology. He gained a Ph.D. from Harvard University in 1895, in history, and had spent a year at the University of Berlin in 1892. Sociology had not emerged as an academic discipline in the 1890s, but DuBois became convinced of the value of a scientific study of society in understanding and then improving the condition of black Americans. According to Rudwick, he 'passionately believed that research could supply the basis for achieving a racially equalitarian [*sic*] society. He contended that race prejudice was caused by ignorance and that social science would provide the knowledge to defeat injustice.' As might be anticipated from DuBois's faith in science, his approach to sociology was heavily empirical: 'I was going to study the facts, any and all facts, concerning the American Negro and his plight and, by measurement and comparison . . . work up to any valid generalization which I could.'[3]

An opportunity to analyse the condition of urban black Americans came in 1896 when the Philadelphia settlement house movement, one of the reformist urban institutions discussed later also in relation to Chicago, invited DuBois to do research on black participation in Philadelphia politics. The outcome of this research was *The Philadelphia Negro*, which, in terms of modern theoretical understanding of the social relations characterizing the capitalist city, is still an impressive study of the social space of the Philadelphia black community. Rudwick[4] describes it as 'a brilliant description of the contours and functioning of the black community, its institutions and its mechanisms for racial survival and advancement'.

THE PHILADELPHIA NEGRO

DuBois worked on this project for fifteen months,[5] focusing on the primary area of black settlement – the seventh ward – where he surveyed almost every household, collecting information on housing conditions, work, literacy and institutions. Much of this information was then represented statistically and cartographically (Figure 8.1). Data gathering was supplemented by participation in community life during the survey period.

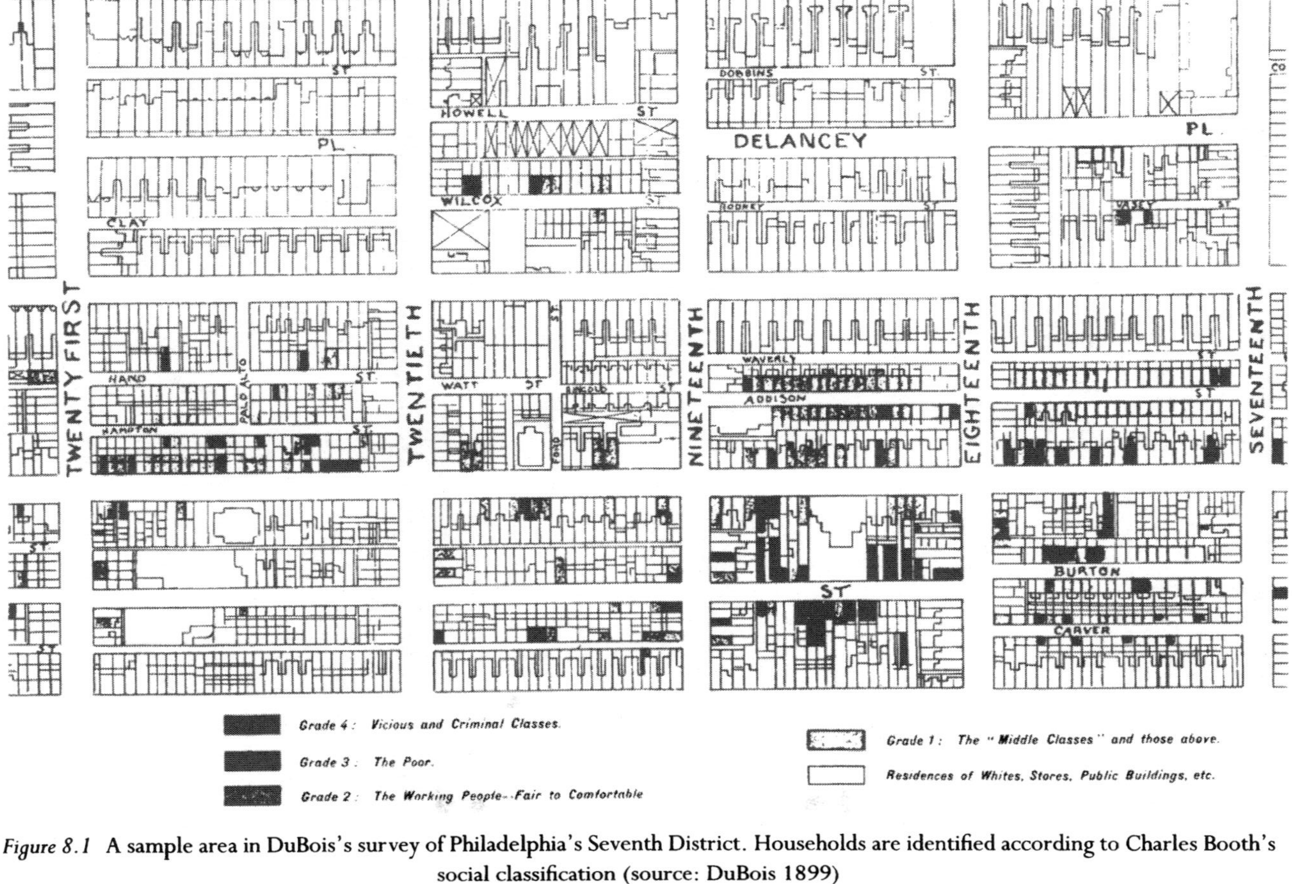

Figure 8.1 A sample area in DuBois's survey of Philadelphia's Seventh District. Households are identified according to Charles Booth's social classification (source: DuBois 1899)

His basic method of research was simple and hard. He knocked at the door of each domicile he could locate in the Seventh Ward, sat down with whomever answered his knock, and asked questions for ten minutes to an hour.[6]

He was aware of both the practical limitations of his survey methods and the inherent subjectivity of his analysis. As he stated in the introduction to the study: 'Convictions on all great matters of human interest one must have to a greater or less degree, and they will enter to some extent into the most cold-blooded scientific research as a disturbing factor' (p. 3).

In categorizing his information, he acknowledged Booth's contribution and employed a similar categorical scheme for describing the black population of Philadelphia to that used by Booth to differentiate the population of London in *Life and Labour of the London Poor*, betraying similar middle-class values.

DuBois's categories were:

Grade 1: The 'Middle Classes' and those above
Grade 2: The Working People – Fair to Comfortable
Grade 3: The Poor
Grade 4: Vicious and Criminal Classes

Elaborating on this classification in the text, he suggests that Grade 4 includes 'the lowest class of criminals, prostitutes and loafers; the submerged tenth'. Rather than being conditioned simply by poverty, DuBois suggests that this group was characterized by 'shrewd laziness, shameless lewdness, cunning crime' (pp. 311–312). The people who comprise Grade 4 are distinguished from the poor in Grade 3, who are unfortunate, unreliable or ignorant. While, from a present-day perspective, the language seems quaint and DuBois's views appear reactionary by comparison with modern views on deviance, elsewhere his analysis shows considerable insight and methodological sophistication. Although he does have a tendency to condemn the poorest of the poor, it would be inappropriate to write DuBois off as a reactionary middle-class observer. As Moore observes, his work was 'the work of a perceptive observer still operating from many of the faulty basic assumptions his culture had provided him with'.[7]

One of the most remarkable features of the study is the treatment of the 'geography' of Philadelphia's black population, which DuBois examines temporally and at different spatial scales. He suggests that 'a study of the Philadelphia Negroes would properly begin in Virginia or Maryland' (p. 75) and proceeds to analyse census data on the birthplaces of blacks in the seventh

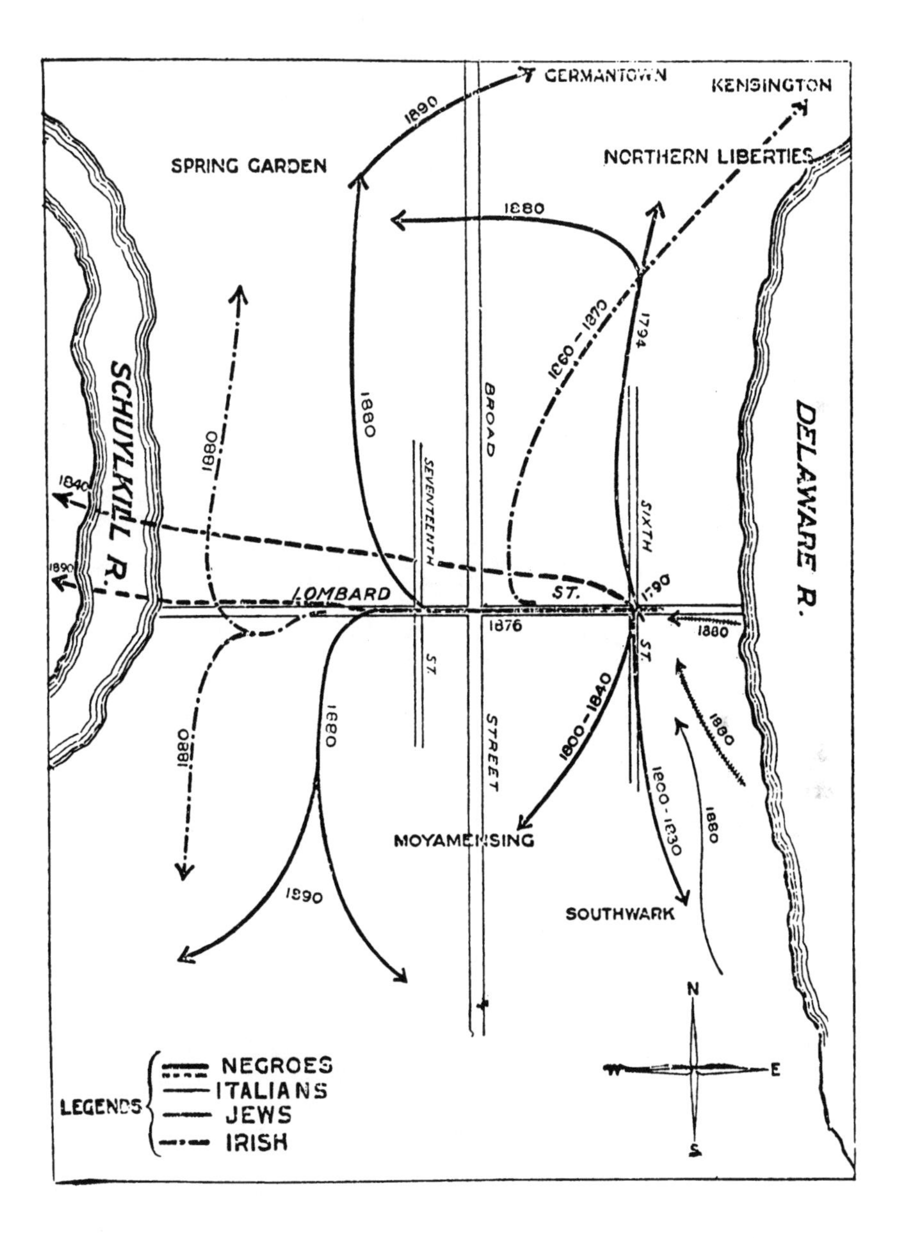

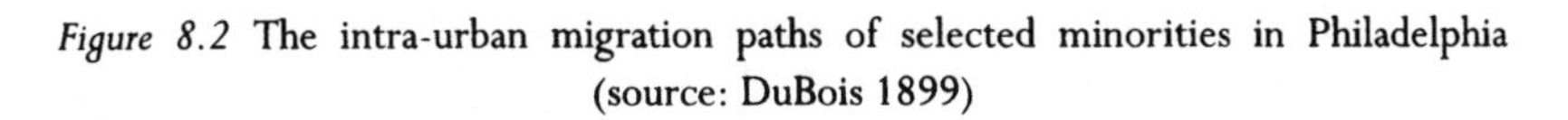

Figure 8.2 The intra-urban migration paths of selected minorities in Philadelphia (source: DuBois 1899)

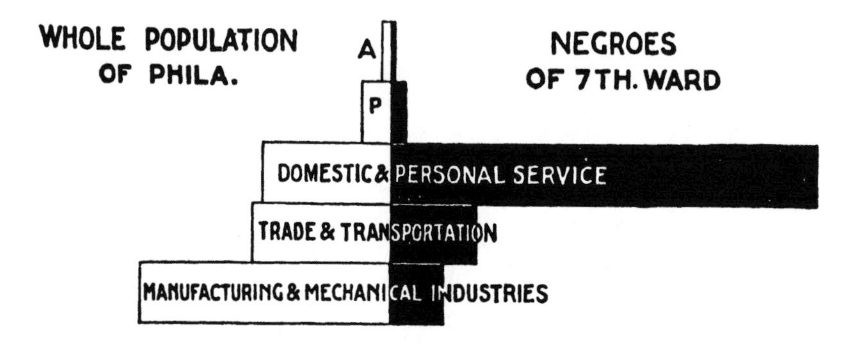

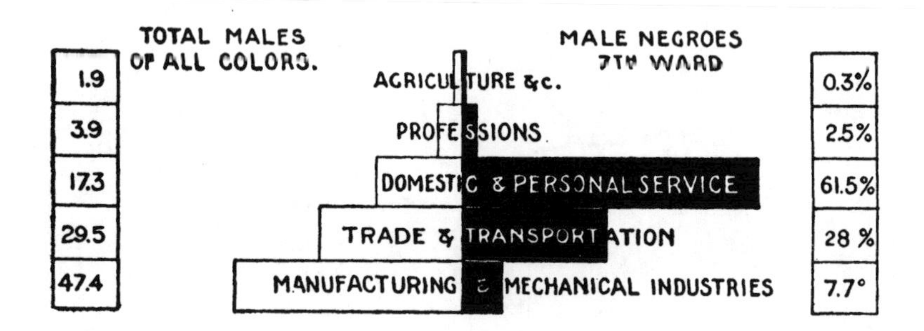

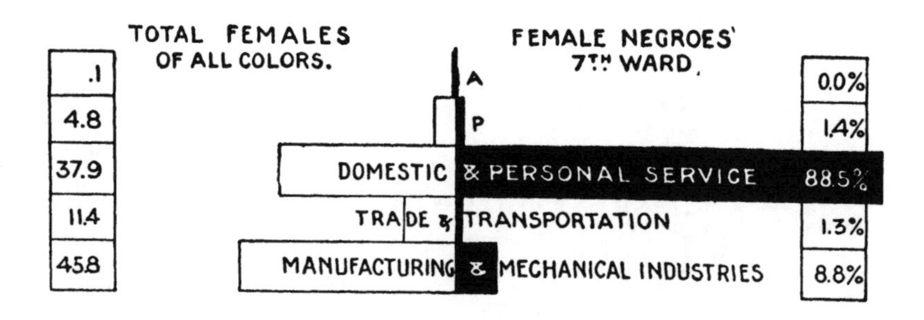

Figure 8.3 Back-to-back histograms used to relate the occupational profiles of black workers in Philadelphia to those of the total working population of the city (source: DuBois 1899)

ward, by age of migrant. DuBois identified a step-wise migration process, with migrants progressing from rural areas to small towns, to larger towns, like Norfolk or Richmond, Virginia, and thence to Baltimore or Philadelphia. (As a part of his larger project, he investigated conditions in the source areas of migrants, work published by the United States Labor Bureau as *The Negroes of Farmville* in 1898.) DuBois describes migration streams within the city, comparing the movement of black and other migrants (Italians, Jews and Irish) between residential areas during the nineteenth century both verbally and cartographically (Figure 8.2). His appreciation of the need for a comparative perspective is evident in a number of analyses, including comparisons of the occupational profiles of the black population in the seventh ward with those of the total population of Philadelphia, of the occupations of black males in the seventh ward with those of all males in the city, and of black females in the seventh ward with all females in the city. These differences are illustrated with back-to-back histograms (Figure 8.3). Using such methods, DuBois effectively situates his study area in the context of the city's economy and social structure. Statistical and graphical description is complemented by detailed case studies which involved interviews and questionnaires, on issues such as crime and job discrimination. Although the methods of data collection and description had been used elsewhere before, this was the first to combine them in a spatial analysis of economic and social processes and, in this respect, *The Philadelphia Negro* has considerable methodological distinction.

DuBois's analyses of residential segregation, discrimination in the workplace, and crime are probably the most important contributions of the study. As Rudwick remarks, 'when *The Philadelphia Negro* was written, [DuBois's generalizations] contrasted strongly with the racist assumptions held by most sociologists and by the general public.'[8] For example, he recognized that the residential patterns of the black population were structured by racism, anticipating modern radical thinking and contrasting markedly with Robert Park's assimilationist view of black minorities in the city.[9] DuBois provided a black perspective which can be distinguished from his more prejudiced comments about the urban poor. DuBois is the 'other', describing the experience of his own community. His awareness of the distinctiveness of the American black population, in terms of its oppression by the dominant white society, is quite clear. He maintains, for example, that

> Many are the misapprehensions and misstatements as to the social environment of Negroes in a great Northern city. Sometimes, it is said, here they are free; they have the same chance as

the Irishman, the Italian, or the Swede; at other times it is said, the environment is such that it is really more oppressive than the situation in southern cities. The student must ignore both of these extreme statements and seek to extract from the complicated mass of facts the tangible evidence of a social atmosphere surrounding Negroes, *which differs from that surrounding most whites*; of a different mental attitude, moral standard, and economic judgement shown toward Negroes than toward most other folk. That such a difference exists . . . few would deny; but just how far it goes and how large a factor it is in the Negro problems, nothing but careful study and measurement can reveal.

(p. 8)

In his account of the housing problem, it is clear that DuBois recognized the existence of housing sub-markets, defined by discriminatory practices and reinforced by the spatial relations of homes and workplaces and the temporal organization of work. Both elements of the problem are identified in the following passage:

Ignorance and carelessness . . . will not explain all or even the greater part of the problem of rent among the Negroes. There are three causes of even greater importance: these are the limited localities where Negroes rent, the peculiar connection of dwelling and occupation, and the social organization of the Negro. The undeniable fact that most Philadelphia white people prefer not to live near Negroes limits the Negro very seriously in his choice of a cheap home. Moreover, real estate agents, knowing the limited supply, usually raise the rent a dollar or two for Negro tenants if they do not refuse them all together. Again, the occupations which the Negro follows, and which at present he is compelled to follow, are of a sort that makes it necessary for him to live near the best portions of the city: the mass of Negroes are, in the economic world, purveyors to the rich – working in private houses, in hotels, large stores, etc. In order to keep this work, they must live nearby. . . With the mass of white workmen, the same necessity of living near work does not hinder them from getting cheap dwellings; the factory is surrounded by cheap cottages, the foundry by long rows of houses and even the white clerk and shop girl can, on account of their hours of labor, afford to live further out in the suburbs than the black porter who opens the store . . . *much of the Negro problem in the city finds adequate explanation when we reflect that here is a people receiving a little lower wages than usual for less desirable work and compelled, in order to do that work, to live in a little less pleasant quarters than most people, and pay for them somewhat higher rents.*

(pp. 295–296)

DuBois documents job discrimination in considerable detail, and this is particularly valuable given his emphasis on the connection between the job market and residential segregation. He makes the point quite forcibly by cataloguing instances of low pay, unfair dismissal and other forms of discrimination against black workers. To give a few examples:

E – was light in complexion and got a job as a driver; he 'kept his hat on' but when they found he was colored, they discharged him.

F – was one of many colored laborers at an ink factory. The heads of the firm died and now whenever a Negro leaves a white man is put in his place.

L – was a first-class baker; he applied for work some time ago near Green Street and was told shortly, 'We don't work no niggers here.'

R – was born in Jamaica; he went to England and worked fifteen years in the Sir Edward Green Economizing Works in Wakefield, Yorkshire. During dull times he emigrated to America, bringing excellent references. He applied for a place as a mechanic in nearly all the large iron working establishments in the city. A locomotive works assured him that his letters were all right, but that their men would not work with Negroes. At a manufactory of railway switches they told him they had no vacancy and he could call again; he called and finally was frankly told that they could not employ Negroes . . . the man has searched for work two years and has not yet found a permanent position. He can only support his family by odd jobs as a common laborer.

(pp. 330–331)

Similarly, DuBois includes a long list of wage differentials to demonstrate the unfair treatment of black workers:

A – got a job formerly held by a white porter; the wages were reduced from $12 to $8.

B – worked for a firm as a china packer and they said he was the best packer they had. He, however, received but $6 a week while the white packers received $12.

C – had been a porter and assistant shipping clerk in an Arch Street store for five years. He receives $6 a week and whites get $8 a week for the same work . . .

and so on.

He concludes that 'the mass of the Negroes have been so often refused openings and discouraged in efforts to better their condition that many of them say, as one said, "I never apply – I know it is useless"' (p. 350).

In accounting for these discriminatory practices, DuBois attaches primary importance to white attitudes or 'public opinion', which sanctions institutional discrimination. Thus,

the object of the trades union is purely business like; it aims to restrict the labor market, just as the manufacturer aims to raise the price of his goods . . . if they could keep out the foreign workman in the same way, they would; but here public opinion within and without their ranks forbids hostile action.

(p. 333)

Apart from connections with the labour market, DuBois recognized practices within the housing market which contributed to residential segregation and created several sub-markets. He suggests that discrimination is

> easily defended from a merely business point of view; public opinion in the city is such that the presence of even a respectable colored family in the block will affect its value for renting or sale; increased rent to Negroes is therefore a sort of insurance, and refusal to rent a device for money-getting.

Then, injecting his own prejudice about class distinctions, DuBois argues that

> Real estate agents also increase prejudice by refusing to discriminate between different classes of Negroes. A quiet Negro family moves into a street. The agent finds no great objection, and allows the next empty house to go to any Negro who applies. This family may disgrace and scandalize the neighborhood and make it harder for decent families to find homes.
>
> (p. 348)

While appreciating the exploitative aspects of residential segregation, DuBois also recognized the advantages of segregation for a black minority within a racist society, anticipating modern arguments against dispersal. He notes the importance of black institutions, particularly churches, in providing support for the community (and requiring a threshold population for their existence) and the isolation and harassment which would be consequences of dispersal. Thus:

> The final reason [for] the concentration of Negroes in certain localities is a social one and particularly strong: the life of the Negroes in the city has for years centered on the Seventh Ward: here are the old churches, St Thomas, Bethel, Central, Shiloh and Wesley; here are the halls of the secret societies; here are the homesteads of the old families. To a race socially ostracized, it means far more to move to remote parts of the city than to those who will in any part of the city easily form congenial acquaintances and new ties. *The Negro who ventures away from the mass of his people and their organized life finds himself alone, shunned and taunted, stared at and made uncomfortable . . . he remains far from friends and the concentrated social life of the church and feels in all its bitterness what it means to be a social outcast.* Consequently, emigration from the ward has gone in groups and centered itself about some church and individual initiative is thus checked. At the same time, color prejudice makes it difficult for groups to find suitable places to move to – one Negro family would be tolerated where six would be objected to; thus, we have here a very decisive hindrance to emigration to the suburbs.
>
> (pp.296–297)

These observations are based on the actual experience of black people; they give an idea of what it is like to be black in a predominantly white society and indicate how social geography is shaped both by white prejudice and social bonds within a black culture. DuBois's analysis is in marked contrast to some more recent studies of urban black populations by white academics which have ignored black perspectives and draw only on the ideas of other whites – the geographical 'self' observing the 'other' as an object. Morrill,[10] for example, was able to argue that residential segregation was bad because it prevented 'normal contacts through which prejudice may gradually be overcome' and advocated clustered dispersal, not because clustering might provide some security for the black migrant, but because this would minimize contact with whites in suburban localities and would not strain the tolerance of white neighbours. This conclusion followed a probabilistic modelling of the black housing market and, in its methodology, the study typified the remote, universalizing view of spatial science which does not engage with people's experience and, thus, has nothing to say about contrasting world-views, and problems like racism. Similarly, a more recent study of migration in Rotterdam describes the residential pattern of black people and foreign workers from Mediterranean countries as if their residential status were a consequence of their cultural background.[11] The authors claim that

> The migration behaviour of these groups is likely to be different from the migration behaviour of the autochthonous groups. For example, they live mostly in the older parts of the city, they tend to inhabit smaller houses with more people, and they usually belong to the lower income groups.

Because the study is concerned with modelling behaviour, it can only work with assumptions about the propensities of migrants. Racism is part of a different vocabulary. DuBois, in contrast to these authors, had much more in common with practitioners employing ethnographic methods, articulating lived experience, than with 'scientific' sociologists and geographers, notwithstanding his own claim that sociology should be statistical and scientific.

Another area in which DuBois made an interesting contribution was in the incidence of crime in the black community. Even though he writes in pejorative terms about the 'criminal class', he does set the problem of crime in an historical and environmental context, identifying the specific negative experiences of black Americans associated with slavery, migration and racism, experiences which may have contributed to criminality. First, he comments in general terms:

> From his earliest advent the Negro, as was natural, has figured largely in the criminal annals of Philadelphia . . . Crime is a phenomenon of organized social life and is the open rebellion of an

individual against his social environment. Naturally, then, if men are suddenly transported from one environment to another the result is lack of harmony with the new physical surroundings, leading to disease and death or modification of physique; lack of harmony with social surroundings leading to crime.

(p. 235)

While this argument may seem simplistic, it again indicates DuBois's awareness of the broader context. Although it is not explicit, it seems that he was arguing that the capitalist city, as well as the historical experience of black people, contributed to crime. At one point, he noted

the immense influence of his peculiar environment on the black Philadelphian; the influence of homes badly situated and badly managed . . . the influence of social surroundings which, by poor laws and inefficient administration, leave the bad to be made worse; the influence of economic exclusion, which admits Negroes only to those parts of the economic world where it is hardest to retain ambition and self-respect.

(p. 285)

Whatever its flaws, *The Philadelphia Negro* provided the first holistic account of urban, black America, and it is one of the very few studies of the urban condition to have been written by a black American. The book has had little impact on modern urban studies, however, and DuBois is little known as an academic writer. Martin Bulmer, who describes the study as a classic, claimed that DuBois 'had almost no influence or imitators'.[12] I will suggest that this neglect is related to the dominating position of the Chicago sociologists in urban studies in the early twentieth century. Blacks, ethnic minorities and women, as students of urban society, were suppressed or marginalized by the dominant white, male centre which comprised the academic establishment.

DUBOIS AND CHICAGO SOCIOLOGY

During the first half of the twentieth century, the University of Chicago and urban sociology appear to be synonymous according to most histories of the subject. Although the most prominent individuals in Chicago's sociology department, particularly Robert Park and Ernest Burgess, were innovative, they naturally drew theoretical inspiration from others, including historians and biologists. However, while the leading figures in the department took it on themselves to inscribe the contours of academic sociology, they were quite selective in the pre-existing knowledge about capitalist societies which they incorporated in the emerging discipline. DuBois's contribution was not a part of this pre-existing knowledge.

Rudwick[13] notes that Burgess and Bogue,[14] in a retrospective account of urban sociology, failed to mention DuBois, although they did make brief reference to Booth's work on London and to the Hull House Papers (discussed in the next chapter). Nathan Glazer, a second-generation Chicago sociologist, acknowledged *The Philadelphia Negro* as an important work in the development of modern sociology, but Burgess and Park, who were trying to give shape to the subject, did not draw on DuBois's writing even though it was close to their main interests.[15]

Park's neglect of DuBois may seem curious because he was one of the few prominent white sociologists in the early twentieth century to involve himself in black issues. His early collaboration with the black political leader Booker T. Washington at the Tuskegee Institute, Alabama, where Park served as Washington's secretary, researcher and ghost writer, was followed by presidency of the Chicago Urban League (1916–1918), a reformist organization concerned particularly with the problems of Chicago's black community. He was also involved in the Illinois Commission on Race Relations investigation into the 1919 race riot in Chicago. Yet, his analysis of race was limited and failed to take account of black history and of DuBois's black perspective in particular. Thus, from his reading of black history, Park argued that: 'Generally speaking, there was no such thing as a race problem before the Civil War and there was at that time very little of what we ordinarily call race prejudice, except in the case of the free Negro.'[16] As Lal observes,

> [Park] sometimes seemed to forget that slavery, discrimination, and segregation stand for relationships which consist of more than just the subjective orientations of groups towards one another. This bias in the direction of social-psychological reductionism – which Park hoped to avoid – also led him to overestimate the role of communication and empathy in mitigating racial injustice.[17]

Park's interpretation of race relations was subject to the mystifications of 'scientific' sociological theorizing in contrast to DuBois's analysis, which immediately identified the central importance of economic exploitation and racism in explaining residential segregation and inequality. Some of Park's comments on residential segregation, derived generally from a scientific view of society and more specifically from free-market models, like filtering theory, demonstrate a singular myopia in regard to the housing opportunities of black Americans. Thus, writing about blacks in Chicago in 1935, he suggested that:

> the migration of the Negroes to northern cities took place at a time when urban residents were abandoning their homes in the centre of the city for the more spacious suburbs. As these suburbs multiplied, the abandoned and so-called blighted areas surrounding the central business core steadily expanded. These areas are now largely occupied by immigrants and Negroes. The result has been, *by a singular turn of fortune*, the southern Negro, lately from the 'sticks' – the man politically farthest down – now finds himself living in the centre of a great metropolitan city [my italics].[18]

Blacks are seen to be progressing up the economic ladder rather than being trapped by discrimination in the housing and job markets.

Fred Matthews attributes Park's limited vision to the influence of Booker T. Washington, although one might wonder why subsequent experience did not lead to a modification of these views. Matthews suggests that

> the attitudes which Washington tried to inculcate as a shield against total despair in the Alabama of 1910 were absorbed by Park and transmitted through him a generation later to an audience of white (and even Negro) social scientists who then preached to a very different audience the 'good news' that political action to protect civil liberties against community mores was futile.[19]

There was, then, a marked difference between Washington and DuBois, the latter espousing the political organization of black people to advance their cause.[20] Matthews goes on to suggest that Park's acceptance of Washington's views on race relations was consistent with the former's conception of social science:

> While the evolutionary view of race relations was widespread early in the century, *Park's espousal of it was vital because of his role in training the experts of the next generation in an increasingly specialized society.* The doctrine produced by the exigencies of survival and political advantage became the scientific sociology of the 1920s and 1930s . . . The unintended effect of Park's teaching, as Gunnar Myrdal pointed out, was to give scientific justification to the Southern racial system as it existed between 1920 and 1940 [my italics].

Bulmer similarly implies that Park disapproved of the politicization of race because political commitment was incompatible with a scientific approach to understanding.[21] Thus, the role of the academic 'was to be that of the calm, detached scientist who investigates race relations with the same objectivity and detachment with which the zoologist dissects the potato bug'.

As Matthews suggests in the passage cited above, because he was in such a powerful position in the profession, Park was able to define the terms of subsequent research on the black population and race relations. He had a considerable influence on E. Franklin Frazier, for example. Frazier, in his doctoral research on the Negro family in Chicago, applied Park's thesis on social disorganization, using terms like 'demoralization' and 'the disorganization of Negro life', and in subsequent writing, for example, *The Negro Family in the United States* and *Race and Cultural Contacts in the Modern World*, his epistemology was essentially Park's. In later work, Frazier did depart from Park's approach, particularly in giving more weight to power relations, but he retained biological concepts, like symbiosis, which were derived from his mentor. According to Edwards, Frazier credited Park with changing the study of race relations 'from a social problems approach to a sociological approach which made for a more objective analysis of behavior in this area'.[22] Appeals to science, particularly to physical analogues, were important in establishing the superiority of sociological as opposed to social survey approaches to investigation, and DuBois could readily be identified with the latter, notwithstanding his own view that his work was scientific.

Park's dominance is similarly evident from accounts of his relationship with Charles Johnson, the black sociologist who developed the research design for a comprehensive study of race relations under the auspices of the Chicago Commission of Race Relations, set up after the 1919 Chicago 'race riot'. Diner maintains that Johnson was supervised 'closely but unofficially' by Park and consulted him frequently during the study.[23] Edward Shils referred to Johnson as Park's 'beloved pupil'[24] and there is little doubt about his concern for black students and black issues. However, Shils indicates the importance of his patronage in the production of research:

> If a graduate student worked on an interesting subject – *usually a subject suggested by him* [Park] – he interested himself unrestrainedly in the student's work. That is how it happened that important studies were published from the dissertations which he supervised by authors who never afterward wrote anything of interest whatsoever. He had no hesitation in telling a student, 'I am interested in you.' He once said that to me.

Given that Park set the agenda for urban sociology during the 1920s and that the Chicago School of Sociology provided models for urban analysis as recently as the 1960s, it is hardly surprising that alternative views failed to register. Black academics apparently had to conform to Park's view of urban society if they were to make any impact, and DuBois's perspective certainly differed in fundamental respects from Park's.

THE FATE OF THE AUTHOR

Following his research in Philadelphia and Virginia, DuBois attempted to establish a research centre for black studies, based in a black college but with close links with Harvard, Columbia, Johns Hopkins and the University of Pennsylvania. He was convinced of the general sociological value of a major study of the African American population, but there was no response from the academic establishment. DuBois moved to Atlanta University in 1897 and attempted to develop the project alone. However, Atlanta was very poorly funded and DuBois could not finance his project from external sources. Nonetheless, between 1897 and 1914 he produced sixteen monographs with the help of part-time, unpaid research assistants. Although these monographs lacked the conceptual clarity of *The Philadelphia Negro*, Rudwick saw them as the beginning of serious sociological research on blacks in the United States. Most of his writing was neglected by contributors to major journals, notably the *American Journal of Sociology*.

In 1909, DuBois became a founder member of the National Association for the Advancement of Colored People and in 1910 he left Atlanta University as a full-time academic. Although he maintained an association with Atlanta and attempted to revive the research proposal in the early 1940s, there was again little interest or finance. Franklin Frazier took over the project at Howard University, but it was abandoned after two years.

Race was not high on the political agenda in the United States until the civil rights movement and the serious urban conflicts in the 1960s. Social issues that are not also national political issues are difficult to promote and it is not surprising that research on America's black population, and W. E. B. DuBois as an academic committed to an understanding of his own people, were marginalized. That DuBois has not been reclaimed by academics working on urban minority issues we might put down to the hegemonic position of the Chicago School of Sociology in urban studies.

CONCLUSION

The neglect of black perpectives in urban sociology applies equally to geography. Geographers have drawn heavily on the Chicago School of Sociology, and Burgess's concentric ring representation of the capitalist city still features prominently in the iconography of the subject. The neglect of a black perpective has resulted in a white view of blackness as 'other', and the perceived 'problems' of black people are essentially problems defined in terms of a white world-view. This is evident in comments by Harold Rose,[25] a black American geographer, writing in 1971, at a time when spatial science provided the paradigm for human geography but dealt inadequately with any questions of cultural difference. He noted that cultural geography had been 'as guilty as other aspects of the discipline in choosing to ignore black America'. He also recognized the importance of subjectivity when he noted that black people might reject externally assigned terms, such as 'ghetto': 'Such a rejection is, indeed, a valid one as a group should have a right to define itself.' (p. 5). The point is underlined in a recent claim by Mitchell and Smith that

> a large segment of the discipline clings to [a] uniracial, unicultural paternalism . . . How else are we to interpret the virtual exclusion of African American, Latino, ethnic Chinese, Native American and other voices from the core of the discipline . . . The entire Milquetoast history of geography as we know and teach it is an exclusionary chronological litany of white, male aristocratic heroes (from Strabo to Humboldt and beyond).[26]

The failure to recognize DuBois's contribution to urban social geography is clearly related to his marginal position within sociology – we tend to refer only to writers in other disciplines who are advertised as authorities. However, an awareness of black perpectives creates a different way of seeing. Both the agency of black people in shaping the city and discriminatory practices which constrain their lives become clearer from reading DuBois. If a search for the marginalized 'other' brings white geographers to the published works of DuBois and other black writers, it will at least contribute to the decentring of the geography of 'the white, male middle', as Mitchell and Smith put it.

To summarize the argument, DuBois probably failed to make an impact, first, because of his methods. He subscribed to a 'scientific' conception of urban sociology which, in the context of the social survey methods current in the 1890s, meant following in the path of Rowntree and Booth. Methodologically, then, DuBois was in the mainstream. His analysis differed from the accounts of urban society offered subsequently by the Chicago sociologists, however. Although scientific in the 1890s, his descriptive social surveys would be judged unscientific by Chicago sociologists in the 1920s, when social science was defined by its use of natural science analogues. DuBois's work did not then fit the dominant paradigm. Second, the neglect of his writing could be related to the emphasis he put on racism in shaping the social geography of

Philadelphia. An interpretation of socio-spatial relations which identified the central role of racism conflicted with the assimilationist views of Park and Burgess, who had the power to marginalize or block alternative perspectives. Thus, conceptions of science and of race can be seen as two *internal* features of DuBois's analysis which contributed to its neglect.

The two are connected. DuBois's experience as a black American encouraged an interpretative or hermeneutic approach to the racial issue and, unlike Frazier later, he was not attracted to natural science models, which could not accommodate fundamental cleavages in society, defined by racism, sexism and class conflict. Natural science provided ideas for an understanding of society in the nineteenth century as well as for pernicious ideologies, particularly those which misapplied Darwinian theory, but it did not become the dominant influence in sociology until the early twentieth century. Moore notes that DuBois always eschewed Social Darwinism, recognizing that it provided a justification for the oppression of black people, and what he termed 'the colored races'.[27] Frazier, working at a time when scientific sociology was being promoted vigorously, appeared to deny his experience as a black American by conforming to Park's models in his early research. An understanding of racism, which requires engagement with minorities, was at odds with the 'objectivity' of scientific sociology.[28]

Apart from these internal factors, concerned with ways of seeing the world, there was also an *external* factor affecting DuBois's marginalization, namely, discrimination in the job market. As a black person, he was unable to work in an institution which had an influential role in the production and dissemination of knowledge. Thus, race was significant both in the formulation of the problem, that is, in suggesting its relevance in understanding the social geography of the city, and in the relegation of a particular analysis, because it was undertaken by a black academic who subsequently found himself in the wrong place.

NOTES

1. Jack Moore, *W. E. B. DuBois*, Twayne Publishers, Boston, 1981. In the late nineteenth and early twentieth century, several prominent black Americans advocated individualism and hard work as a route to equality with whites. For example, *Co-operation*, a weekly publication of the Chicago Bureau of Charities which appeared between 1900 and 1908, included articles by Booker T. Washington, who consistently argued the case for self-help. Similarly, in one issue,

a Chicago minister advised that the black man should acquire property so that 'his dual position, that of man of character plus man of property, will command respect and honor from whites'. This trust in the capitalist system and 'the American way' was shared by DuBois, but it did not blind him to prejudice and discrimination. He became one of several middle-class blacks who were active in campaigning for equal rights through the NAACP and organizations like the Chicago Urban League.

2. There are exceptions. Some white geographers have demonstrated a commitment to the understanding of black experience, including Bill Bunge, in *Fitzgerald: Geography of a revolution*, Schenkman, Cambridge, Mass., 1971, and Peter Jackson in his *Maps of Meaning: An introduction to cultural geography*, Unwin Hyman, London, 1989, and other work. It is still difficult, however, for white people to articulate black world-views, and it is better if black people do it themselves.

3. Cited by Elliott Rudwick, 'W. E. B. DuBois as sociologist', in James Blackwell and Morris Janowitz (eds), *Black Sociologists: Historical and contemporary perspectives*, Chicago University Press, Chicago, 1974, pp. 25–55.

4. ibid., p. 28.

5. *The Philadelphia Negro: A social study*. The book was first published in 1899. Page references are to the 1967 edition, University of Pennsylvania Series in Political Economy and Public Law 14, Benjamin Blom, New York.

6. Moore, op. cit., p. 38.

7. ibid., p. 49.

8. Rudwick, op. cit., p. 33.

9. Peter Kivisto, 'The transplanted then and now: the reorientation of immigration studies from the Chicago School to the new social history', *Ethnic and Racial Studies*, 13 (4), 1990, 455–481.

10. Richard Morrill, 'The Negro ghetto: problems and alternatives', *Geographical Review*, 55, 1965, 339–361.

11. D. Ament and G. van der Knapp, 'A two-stage model for the analysis of intra-urban mobility processes', *Environment and Planning A*, 17, 1985, 1201–1216.

12. Martin Bulmer, *The Chicago School of Sociology*, Chicago University Press, Chicago, 1984, p. 66.

13. Rudwick, op. cit., p. 25.

14. Ernest Burgess and Donald Bogue, *Contributions to Urban Sociology*, Chicago University Press, Chicago, 1964.

15. There were several references to Dubois in Park and Burgess's *An Introduction to the Science of Sociology*, Chicago University Press, Chicago, 1921, but no extracts from his work. I am grateful to Peter Jackson for this information.

16. Barbara Lal, *The Romance of Culture in an Urban Civilization*, Routledge, London, 1990, p. 133.

17. ibid., p. 63.

18. ibid., p. 142.

19. Fred Matthews, *Robert E. Park and the Chicago School*, McGill–Queen's University Press, Montreal, 1977, pp. 81–82.

20. The political differences between Washington and Dubois are discussed in A. Spear, *Black Chicago: The making of a Negro ghetto, 1890–1920*, Chicago University Press, Chicago, 1967.

21. Bulmer, op. cit., p. 76.

22. Franklin Edwards, 'E. Franklin Frazier', in Blackwell and Janowitz, op. cit., p. 111.

23. Steven Diner, *A City and its Universities: Public policy in Chicago, 1892–1919*, University of North Carolina Press, Chapel Hill, 1980, p. 132.

24. Edward Shils, 'Some academics, mainly in Chicago', *American Scholar*, 50, 1980, 179–196.

25. Harold Rose, *The Black Ghetto: A spatial behavioral perspective*, McGraw-Hill, New York, 1971, p. 5.

26. Olivia Mitchell and Neil Smith, 'Bringing in race', *Professional Geographer*, 42 (2), 1990, 232–234.

27. Moore, op. cit., p. 42.

28. Martin Bulmer suggested that the detachment and objectivity of Park and his black students, particularly Charles Johnson, were virtues. He claims, for example, that

> The lasting value of the report [of the Chicago Commission on Race Relations] was . . . its demonstration of the relevance of social science to policy issues. This was achieved not only by comprehensive inquiry and empirical documentation, much of it novel by the social science standards of the time, but also by a very high degree of scientific detachment from the emotional subject matter it was dealing with. Both Johnson and Park developed this detachment to a very high degree.
>
> (Martin Bulmer, 'Charles S. Johnson, Robert E. Park, and the research methods of the Chicago Commission on Race Relations, 1919–1922: an early experiment in applied social research', *Ethnic and Racial Studies*, 4 (3), 1981, 289–306)

9

RADICAL WOMEN, MEN OF SCIENCE AND URBAN SOCIETY

It would be naive to claim that *The Philadelphia Negro* was neglected solely because its author was black. Similarly, in this chapter, which is concerned with urban geographies produced by women in the School of Social Service Administration in the University of Chicago, I do not argue that gender alone accounts for their eclipse.[1] Rather, I will suggest that gender, politics and epistemology combined to marginalize women who were doing significant research on the city, specifically in Chicago between about 1910 and 1930. The time and place are important because there was a particular conjunction of politics, social relations and locale with which the writing of a group of female academics can be associated. The context has to be specified to make sense of their analyses. I would contrast this approach to understanding a particular set of ideas with the usual presentations of that partial view of the city articulated by Park, Burgess and colleagues in the Chicago School of Sociology. Thus, in spatial science texts, like Haggett's *Locational Analysis in Human Geography*,[2] or in Schnore's 'On the spatial structure of cities in the two Americas',[3] Burgess's socio-spatial model of the city is torn out of its Chicago context and discussed with little criticism, apart from some reservations about its spatial application in other unspecified locales.

Such knowledge, stripped of its ideological and historical content, is, as Haraway suggests, 'unlocatable and so irresponsible'.[4] It is irresponsible because it cannot be called into account. The authors' politics are hidden and, in the case of Burgess's model, the geometry is place-less. The relevance of the world-view of the authors to their portrayal of the city is denied and cannot be scrutinized. Thus, my intention in the following account of urban space and social groups as they were interpreted by women in early twentieth-century Chicago is to establish the connection between the world-view of the authors and their portrayals of urban society and then to contrast the accounts of the city by Chicago sociologists like Robert Park – accounts which had a white, middle-class signature.

If we are to explain the relative obscurity of the Chicago women, it is necessary to look both at the content of their analyses and the response of the male-dominated university establishment, to connect politics, prejudice and academic knowledge. In part, this will involve assessing poorly documented relations between individual women and men and social practices both inside and outside the academy. Before attempting to reconstruct this history, however, I want to pursue the idea of gendered knowledge. Can a distinctive body of ideas be associated with women, and why would 'women's knowledge' appear threatening to men?

GENDER AND SCIENTIFIC KNOWLEDGE

Scientific knowledge, rather than other kinds of knowledge, is relevant to this argument because of the elevated status given to science by the Chicago sociologists. Knowledge which was unscientific, according to Robert Park in particular, was downgraded in a hierarchy of knowledges. There might, of course, be a case for assessing the value of science without reference to politics or ideology. As Evelyn Fox Keller observes: 'As long as the course of scientific thought was judged to be exclusively determined by its own logical and empirical necessities, there could be no place for any signature, male or otherwise, in that system of knowledge.'[5] However, science, as a part of the hegemonic culture, is instrumental in maintaining power and domination by men. Objectivity and universality may be seen not just as characteristic of science but as masculine values in that they contribute to the domination of women. 'Universal truths' may be masculine values in a coded form. This, of course, conflicts with the commonly held view of science as emotionally and sexually neutral,[6] but the equation of science and objectivity, objectivity and masculinity has been argued by a number of modern feminist writers and, earlier, by the sociologist George Simmel.

An explanation of such a link between science and masculinity has been suggested by Keller, drawing on object relations theory. She maintains that the maternal environment of infancy, coupled with the cultural definition of masculine (that which can never appear feminine) and of autonomy (that which can never be compromised by dependency) leads to the association of female with the pleasures and dangers of merging, and of male with the comfort and loneliness of separateness.[7]

There is no conception of separateness (or autonomy) in the early stages of infancy – there are no boundaries separating the child's internal and external environments and the 'external environment, consisting primarily of the mother during this early period, is experienced as an extension of the child'. An awareness of gender difference, however, comes with the separation of self and other:

The basic and fundamental fact that it is, for most of us, our mothers who provide the emotional context out of which we forge the discrimination between self and other inevitably leads to a skewing of our perceptions of gender. As long as our earliest and most compelling experiences of merging have their origin in the mother–child relation, it appears to be inevitable that the experience will tend to be identified with 'mother', while delineation and separation are experienced as a negation of 'mother', as 'not-mother'.

Because boys develop their sexual identity 'on an opposition to what is both experienced and defined as feminine', gender identity accentuates the processes of separation and autonomy. Keller recognizes that the separation of self and other is reinforced by the dominant culture, which equates masculinity with autonomy and femininity with merging. Thus, '*the boy's internal anxiety about both self and gender is echoed by the more widespread cultural anxiety, thereby encouraging postures of autonomy and masculinity*'. Crucially, she goes on to suggest that the separateness and autonomy of the boy, the separation of the male self from others (including both subjects and objects), results in objectivity in itself becoming associated with masculinity. Then, cultural pressures lead to 'the entrenchment of an objectivist ideology and a correlative devaluation of (female) subjectivity'. What Keller presents is an argument which suggests an

interaction between gender development, a belief system which equates objectivity with masculinity, and a set of cultural values which simultaneously elevates what is defined as scientific and what is defined as masculine. The structure of this network is such as to perpetuate and exacerbate distortions in *any* of its parts – including the acquisition of gender identity.

Like Keller, Iris Young maintains that objectivity itself comes to be associated with masculinity.[8] Young is then able to characterize a masculine science, an enterprise which reduces 'the self to pure mind, abstracted from sensuality and material immersion in nature'. Pursuing this path, we might argue that the structuring of academic knowledge into 'pure' subjects and the separation of academic knowledge from practice are also masculine traits, reflecting an abhorrence of mixing or merging. Mixed knowledge could be seen as dangerous knowledge, knowledge which threatens hierarchical power structures.

Autonomy, separateness, while comforting, are also painful and object relations theorists have argued that an impulse to dominate others is a natural concomitant of male autonomy. The male child tries to master the 'other' and 'the original self-assertion is . . . converted from innocent mastery to mastery over and against the other'.[9] Feelings of security and well-being come from domination, which compensates for the loneliness and pain of separateness. Thus, the objectivist ideology of the scientific method (denying the inherent subjectivity of science) and the urge to dominate are suggested as *linked* masculine traits. This does not mean, of course, that women cannot subscribe to the objectivity and universality of science. Nor does it mean that science is necessarily an instrument of domination and oppression. It can be liberating. Also, the science referred to here is 'western' science. Science which derives from a Buddhist tradition, for example, constitutes a different system of knowledge. The message of object relations theory, however, is that 'western' science embodies masculine values of objectivity, detachment and universality which are culturally produced and reproduced as part of the distinction drawn between male and female in childhood. Scientific ideology then serves to maintain male domination.

This argument might appear unduly essentialist. The same argument can be presented in different terms. For example, the characterization of masculine or masculinized knowledge is similar to the defining characteristics of truth according to some ancient Greek philosophers whom Feyerabend cites in arguing that modern disputes in the philosophy of science had venerable antecedents.[10] The debates among the Greek philosophers, which expressed an opposition between scientific and humanist positions, was pursued without reference to gender. I would argue, however, that in certain contexts questions of knowledge can be usefully addressed in terms of gender. The case that I consider in this chapter is one in which the idea of gendered knowledge seems entirely apposite.

GENDER, POLITICS AND KNOWLEDGE

The assertion that knowledge can have a male or female signature gains support from accounts of social science in Chicago in the early twentieth century, but I first want to suggest that Keller's argument has some generality by referring to another study of scientific knowledge, by Potter,[11] a study which also has relevance to the case of black social scientists. Potter was interested in a seventeenth-century debate on the nature of matter, involving the chemist Robert Boyle and radicals pressing for reform or revolution during the English Civil War. This may seem highly esoteric and remote from the question of sociological knowledge in Chicago, but the case does demonstrate how politics, particularly gender politics, and science can interrelate. The parallel, in fact, is quite a close one.

In England in the 1640s, there were several radical movements with egalitarian programmes which were never realized, notably the Diggers and the Levellers. The politics of the Diggers was informed by a philosophy which Potter calls 'a materialist pantheism', expressed in Gerrard Winstanley's statement that 'To know the secrets of nature is to know the works of God.' This belief is part of a natural magic tradition, according to which all matter is alive and in movement (hylozooism). Such ideas were current among some mediaeval scholars in Europe, and they were taken up by the sixteenth-century physician Paracelsus, who argued on the basis of this belief for social equality – if the spirit is in everything and in everybody, there is a clear logic in campaigning for human equality. Hylozooism and materialist pantheism inspired revolutionary ideals. Thus, according to the Leveller John Lilburne:

> Every particular and individual man and woman that ever breathed in the world since [Adam and Eve] are and were by nature equal and alike in power, dignity, authority and majesty, none of them having (by nature) any authority, dominion or magisterial power, one over . . . another.
>
> (Potter, p. 23)

Robert Boyle's early writing was strongly influenced by the same philosophy. He drew on hermeticism,[12] alchemy and natural magic, and he identified with Paracelsus and others working in the same tradition. However, Boyle came to recognize the revolutionary implications of natural magic and repudiated the philosophy at a time when the revolutionary threat was becoming a real one, in the late 1640s. In a radical shift of position, he advanced an atomistic or corpuscular theory of matter, according to which matter was dead, inert. As Potter recognized:

> As long as God or the World Soul inhabits even the atoms, it inhabits all men and allows them to know all that they need to of spiritual matters without priestly mediation [which served to reinforce hierarchies and keep people in their place]. Refutation of the sectaries' [revolutionary movements'] natural and social philosophies demands, then, at least that matter be passive, dead, brute or inert.

Thus, through his corpuscular theory, Boyle distanced himself from the Diggers and the Levellers, who were threatening his class interests.

Potter demonstrates that the opposition of material pantheism and corpuscular theory had particular implications for women. In the 1640s, Boyle wrote a number of essays on women in which he lent support to the establishment view that a woman's place was in the home, that they should not own property or engage in political activity. At this time, women, including Leveller women, were politically active, demonstrating against the Civil War, high food prices and taxes and demanding the release of Leveller men from prison.

> Sound Englishmen were quick to recognize in the new sectarian views – that all people are equal, and in particular, that women are at least spiritually equal to men *since every one alike has the spark of life within them* – a serious threat to the sexual status quo.
>
> (Potter, p. 26)

Potter's essay is important because it demonstrates that a world-view which informs politics and programmes for social change can also inform science, which, in turn, informs politics. Rather than science constituting a system of knowledge which is judged solely according to its internal logic, empirical validity and predictive power, it can be seen, in its internal constitution, as knowledge which is imbedded in systems of belief. Thus, science can be used to further political interests, including the domination of women by men. I will now develop this argument with reference to the women in Chicago who stood in opposition, intellectually and politically, to the men of the Chicago School of Sociology.

JANE ADDAMS, HULL HOUSE, SOCIALISM AND GEOGRAPHY

Central to this part of my argument are an obscure text, *The Tenements of Chicago*,[13] produced by the staff and students of the Chicago School of Social Service Administration in the 1930s, and fragmented accounts of the relationship between some of the book's authors and the men of the Chicago School of Sociology. If we are to understand the response to the book or, rather, the almost total lack of response by contemporaries and modern urban theorists, the writers have to be put into a context defined by national, local and university politics, by gender relations, by time and place – Chicago in the early twentieth century.

Setting the scene first involves assessing the influence of one key figure, Jane Addams, whose intellectual and political concerns ranged from the local to the global. Her global range was expressed in an international network of intellectuals and practitioners which developed around the principal node of the Hull House settlement in Chicago. Hull House was important both as a forum for debates on the city, American politics and global issues and as a source of practical assistance for immigrants. Most of the women who undertook research on Chicago's housing and minority problems in the 1920s and 1930s, including Edith Abbott, who initiated much of this research, spent some time as residents of Hull House and were strongly influenced by Jane Addams. The perspectives on urban society developed by Hull House residents and the politics of Jane Addams are closely linked.

The Hull House settlement was founded in 1889 by Jane Addams and Ellen Starr. It was located at the intersection of Halstead and Polk Streets, in one of the main immigrant reception areas of the city. In its original conception, Hull House was designed to bring together the local working-class population and middle-class academics and social workers. As one of the early residents, probably Addams, envisaged the settlement:

> This centre . . . to be effective, must contain an element of permanency so that the neighborhood may feel that the interest and fortunes of the residents are identical with their own. The settlement must have an enthusiasm for the possibilities of its locality.[14]

She continued:

> The original residents came to Hull House with a conviction that social intercourse could best express the growing sense of the economic unity of society. They wished the social spirit to be

the undercurrent of the life of Hull House, whatever direction the stream might take. All the details were left for the demands of the neighborhood to determine, and each department has grown from a discovery made through natural and reciprocal relations.[15]

This concern for understanding the life-worlds and problems affecting recently arrived immigrants in the city, documented in *Hull House Maps and Papers* (published in 1895), was accompanied by an increasing political awareness, particularly in regard to the exploitation of the immigrant working class and women, and for Jane Addams, this became an essentially socialist perspective. As Lasch suggests:

Even those like Jane Addams who did not embrace socialism, and whose political position therefore has to be described, for the lack of a better word, as 'progressive' (or 'liberal'), had more in common with socialists than with the kind of progressives one associates with . . . the crusade for 'good government', etc. What distinguished her from them was not only her insistence on the preeminence of 'education' but her sense of kinship with the 'other half' of humanity. The intellectual in his [*sic*] estrangement from the middle class identified himself with other outcasts and tried to look at the world from their point of view.[16]

A number of prominent intellectuals associated themselves with this view, including the pragmatist philosopher John Dewey. According to Davis:

Dewey himself repeatedly used Hull House and a few other settlements as models for what he hoped schools would become, but it was a reciprocal relationship and Addams and the other residents learned a lot from him. He made them realize the implications of their programs and experiments.[17]

Hull House developed a political perspective with both local and international dimensions. At a local level, opposition to the state was expressed in the role of Hull House as a refuge for immigrants threatened with deportation. At the same time, residents exchanged ideas with European socialists, including the British Labour leader, Keir Hardie, and the Fabian socialists Sidney and Beatrice Webb, and the anarchist Kropotkin. Kropotkin's visit to Hull House, to talk about his book *Factories, Fields and Workshops*, led to subsequent allegations that the place was a base for anarchist subversion, particularly after the assassination of President McKinley, when there was an anarchist panic in Chicago.[18] While these political figures were attracted to Hull House, Jane Addams also travelled extensively and became involved in international issues. In 1896, she had meetings in London with the leaders of the new London County Council, with the dockers' leader, Ben Tillet, who rowed her down the Thames, with Octavia Hill, the housing reformer, and the Webbs. She also visited the Bermondsey settlement. Addams then travelled to Russia and, via Nijni Novgorod and Moscow, visited Tolstoy at his home in Yasnaya Polyana. Tolstoy clearly made a great impression on her. In *Twenty Years at Hull House*, she wrote that

> Tolstoy has made the one supreme personal effort, one might almost say the one frantic personal effort, to put himself in the right relations with the humblest people, with the men who tilled his soil, blacked his boots and cleaned his stables.
>
> (pp. 267–70)

She did not suggest, however, that he black his own boots. Some of Addams's close associates had a more radical outlook. Florence Kelly, for example, another founder member of Hull House, worked with the Socialist Party in New York City and in her book *Modern Industry* maintained that the only real solution to industrial problems was cooperative ownership of the means of production.

A clearer political position is evident later in Addams's comments on women's suffrage, following her involvement with the International Suffrage Alliance in 1913. Attending a conference of the Alliance in Budapest, she stressed the importance of women's solidarity and internationalism.[19] Later, after a women's conference in Honolulu in 1928, she perceptively noted the broader effects of western imperialism on east and south-east Asia:

> We have come to regard the advance of western civilization as . . . an effort to make one type of culture predominate which, if too hardly pushed, may break down other cultures, older and more basic. We have been accustomed to say with pride that telephone poles may be set up in the jungle, that a wireless operates as easily on sea as on land, forgetting that in our absorption in communicating culture, we may easily lose the patterns and customs which give it any real value.

Translating these ideas into political action, Jane Addams was involved with Sophonisba Breckinridge in the organization of a series of Women's World's Fairs, designed to increase awareness of the international women's movement.[20]

Opposition to war was another aspect of Addams's politics, and a particularly important one in relation to the negative response of the male establishment to Addams and her associates, as academics and practitioners. During the First World War, she joined the Women's Peace Party, which had close links with the Organisation Centrale pour une Paix Durable in The

Hague, and the League for Democratic Control in England. In 1916, Jane Addams, Grace Abbott and Sophonisba Breckinridge, all at the time resident at Hull House, attended the International Congress of Women Against the War, in The Hague. Deegan suggests that Addams's pacifism

> drove a further wedge between her and the men of the Chicago School [of Sociology]. Thomas, who was closest to Addams . . . and even Dewey and Mead, differed from her concerning the war. Her position, which had always been strongly identified with women, was increasingly defined as oppositional to the men's.[21]

Elsewhere, Emily Greene Balch, a close associate of Addams and a founder of the Denison settlement house in Boston, lost her job as professor of sociology and economics at Wellesley College because of her opposition to the war. Both women were later to win the Nobel Peace Prize, Addams in 1931 and Balch in 1946.[22]

What Addams's detractors described as 'internationalism', a term coupled with pacifism and socialism in right-wing demonology, was evident also in her response to xenophobia in Chicago during the war:

> Chicago, with its diversified population, inevitably displayed many symptoms of an inflamed nationalism, perhaps the most conspicuous were the deportations and trials of 'Reds'. Throughout the period of the war, we were very anxious that Hull House should afford such refuge as was legitimate to harassed immigrants . . . I was thankful that we got through the entire period of the war and post-war without a single arrest at Hull House, if only because it gave a certain refuge to those who were surrounded by the suspicions and animosities inevitably engendered by the war toward all aliens.[23]

In the decade after the First World War, nationalist and xenophobic sentiments were very strong in the United States and, in an atmosphere comparable to that during the witch hunts of Senator Joe McCarthy during the 1950s, Jane Addams was singled out as a subversive. In 1919, she headed a list produced at a Senate sub-committee of sixty-two persons who held 'dangerous, destructive and anarchistic sentiments'.[24] In 1920, a four-volume study of 'revolutionary radicalism' by Senator Clayton R. Lusk implicated Addams: 'The technique was simple; masses of real documents and accurate information were printed with completely false and half-true statements of connections with international communism, Jane Addams's name appeared frequently in the four volumes.'[25] At a meeting in Chicago in 1920 she attacked the Federal government for its reactionary policies, arguing that:

> The government is proceeding on the theory that because these thinking aliens demand an end

of the class struggle and equal rights for all, they are plotting to overthrow the United States. So it was said of suffrage years ago. Anything that is radically new to the established order of things is revolution in the eyes of the many.

One headline in a Chicago newspaper the next day declared that 'Jane Addams Favors Reds.' This abuse continued throughout the 1920s. For example, there were several versions of a purported network of communist organizations, implicating Jane Addams and the whole Hull House community. Similarly, an article in *The Woman's Patriot* in 1926 suggested that 'practically all the radicalism [i.e communism] started among women in the United States centres about Hull House, Chicago, and the Children's Bureau at Washington' (where Grace Abbott, a former Hull House resident, was director). More extravagantly, a book called *The Red Network* (1934) claimed that 'the influence of [Jane Addams's] radical protegées, who consider Hull House their home center, reaches out all over the world'.

SOCIAL WORK, SOCIAL SCIENCE AND SEXISM

This communist, subversive taint is compounded by other prejudices against Jane Addams and her associates, particularly concerning gender and the 'scientific' status of their academic work. A connection between gender and knowledge is apparent in the sense that much of the group's research was dismissed as 'women's knowledge' by Robert Park and others. However, to describe the negative response to their writing on urban society as 'sexist' oversimplifies. Their politics was also a factor and, while some political attitudes derived from their position as women in a patriarchal society and might be described as feminist politics, they clearly shared other socialist ideals with men, like Upton Sinclair, for example. As women who were politically to the left, they suffered double exclusion.

In criticisms of Jane Addams and, more particularly, Edith Abbott and her co-workers who were working on urban population and housing questions at the same time as Park and Burgess, there were implicit, and occasionally explicit, associations made between gender and knowledge. Sociology, as it was represented by Park, was hard, scientific and masculine, whereas social work, almost a preserve of women at the University of Chicago, was unscientific and feminine. The scientific discipline of sociology was judged to be superior to, and more worthwhile than, social work. This judgement is suggested in comments by Queen reminiscing about his time as a student at the University of Chicago from 1910 to 1913, and in 1919:

> I was immersed in sociology as a 'science' and, except in the classes of Charles R. Henderson and Edith Abbott, I heard little about the social services. However, there were occasional rumblings about 'the old maids downtown who were wet-nursing social reformers'[26]

(a reference to Edith Abbott and Sophonisba Breckinridge in what was then the Chicago School of Civics and Philanthropy). As single, professional women, they did not conform to the patriarchal image of woman as mother, but they were cast in this role by their critics, as mothers of social workers. According to Martin Bulmer,[27] Park was forthright in his criticism, dismissing the women as academics and attacking their politics. Thus,

> One student recalled Park telling a seminar that the greatest damage done to the city of Chicago was not the product of corrupt politicians or criminals but the women reformers. Park's personal relations with Misses Abbott and Breckinridge were distant and he discouraged students from taking their courses.

The persistence of this distinction between 'scientific' sociology and social work at the University of Chicago, between a male sphere and a female sphere, is suggested by Peter Jackson's observation that:

> The School of Social Service Administration [formerly the School of Civics and Philanthropy] is still symbolically marginalized in Chicago, situated across the Midway from the main campus of the university. It is still regarded by sociologists as a rather inferior place, dedicated to the training of professional social workers rather than the advancement of scientific sociology, an implicitly masculine form of knowledge . . .
>
> (personal communication)

It should be noted, however, that not all male sociologists in Chicago shared Park's prejudices. Albion Small, the founder of the sociology department, had considerable sympathy with Jane Addams's approach to social reform, and Ernest Burgess argued that sociologists would benefit from closer collaboration with social workers, and he had particular admiration for Breckinridge's work.[28]

The importance of a 'scientific approach' as a defining characteristic of real sociological knowledge is stressed repeatedly by the Chicago sociologists[29] and this clearly contributed to the sidelining of research done under the heading of social work. This distinction between social work in the field and scientific studies which were appropriately housed in universities also affected some of Abbott's colleagues outside Chicago. She reflected that 'when we decided in Chicago to turn our old school over to the university, we were severely criticized by our eastern colleagues who told us we could not have casework and field work in a university'.[30] Even Burgess, while arguing for closer links between sociology and social work, saw science as necessary for an understanding of social issues, and ignored urban research which did not fit the dominant paradigm. Thus, while acknowledging the Hull House papers as the beginning of systematic urban studies (no mention here of W. E. B. DuBois), he still maintained that it was a scientific sociology, as developed by Park and his associates, that provided an understanding of social problems 'in terms of the processes and forces that produce them'.[31]

To understand their distinction between sociology and social work, we have to take account of Park's education, and particularly his period of study in Germany, where he was strongly influenced by Windelband, Rickert and Simmel. Park recalled that

> Windelband described philosophy as a 'science of sciences', fundamentally a science of method based on a history of systematic thought. There is, in my opinion, no other way of getting an adequate conception of scientific method. I learned a lot from Windelband.[32]

As Entrikin has pointed out, Windelband and Rickert initially made a distinction between nomothetic and idiographic sciences but they subsequently moderated their positions, acknowledging that sciences combined elements of the universal and the particular.[33] Park, however, stuck with the earlier distinction. Accepting Comte's use of the abstract and the concrete, he argued that human ecology was an abstract science, making use of the facts of geography – a concrete science – to develop theory through the application of the scientific method. This is analogous to the distinction drawn by Park and Burgess between sociology (abstract) and social work (concrete), as Burgess had suggested in his 1927 paper.

Apart from the philosophical justification for the position taken by Park and Burgess, scientific sociology appeared also apolitical, and it provided support for the prevailing ideology in the United States in the 1920s. Deegan suggests that the First World War brought an end to idealism in America and provided a fertile ground for Freudian ideas that attributed social action to individual drives and conflicts.[34] I would suggest instead that it was the growing strength of American capitalism, coupled with nativist and xenophobic attitudes, which provided a fertile ground for political philosophies which favoured individualism over collective action, and a positivist, scientific sociology was consistent with such political perspectives. Park's sociology has continued to strike a sympathetic chord in the conservative academy. Retrospectively, it can be appreciated that the urban research of Abbott, Breckinridge and other women who combined theory and practice in a political project did not constitute 'legitimate knowledge' until Marxist and Weberian analyses gained a foothold in urban studies, by which time their contribution had been lost from sight.

THE SCHOOL OF SOCIAL SERVICE ADMINISTRATION AND THE CHICAGO STUDIES

The research done in this school of the University of Chicago continued the investigations into urban poverty, inequality and ethnicity initiated by Florence Kelly and Jane Addams at Hull House in the 1890s and published as *Hull House Maps and Papers* in 1895. The central figures in the School, which was founded in 1924, were Edith Abbott and Sophonisba Breckinridge, but there were other research workers, largely unacknowledged, who made several remarkable contributions to the Chicago project.

Edith Abbott had a background which, in several respects, paralleled that of Jane Addams. Edith and her sister Grace came from a professional middle-class family in Grand Island, Nebraska. Their mother, like Jane Addams, had a concern for the oppressed; she was a pacifist and committed to equal rights for women.[35] Edith Abbott received a doctorate from the University of Chicago in 1905 and then worked in Boston as secretary for a women's trade union league and as a researcher for the American Economic Association, where she was involved in projects concerned with wages, prices and women's work. The latter involved investigating women's employment in the cigar and garment-making industries in New York City. This research led to a book, *Women in Industry*, published in 1910, and a series of articles in the *Journal of Political Economy* and the *American Journal of Sociology*. In 1906, she had a year in London, studying at the London School of Economics with Sidney and Beatrice Webb, and, on her own initiative, visiting settlement houses in London and conducting surveys and interviews with the poor in the East End, from a base at St Hilda's settlement. This experience led her to the conclusion that the Fabian socialists, like the Webbs, were elitist and lacked a real understanding of the problems of the poor. After one year teaching at Wellesley College, Edith and Grace Abbott started a ten-year stay at Hull House (1908–1918), a period when social conflicts were acute, but also a period of considerable intellectual stimulation because this was the time when the settlement was a major centre of radical thought. Subsequently, Grace was able to apply her ideas on social policy as Director of the Immigrants' Protective League and then as Director of the Children's Bureau in Washington, D.C. Edith resumed her academic career in the School of Civics and Philanthropy and, later, as Dean of the School of Social Service Administration. It is evident from acknowledgements in books, references and biographies that the Abbotts were a part of a closely connected network of women with a shared view of capitalist society. It was possible for detractors to refer to the women collectively, to see them as a group who were critical of American society and were in this sense un-American, as well as being intellectually impressive and a challenge to men within the academy.

THE TENEMENTS OF CHICAGO

The research project which provided the material for *The Tenements of Chicago, 1908–1935,* published by the University of Chicago Press in 1936, was directed by Edith Abbott and based at Hull House. In the preface, Abbott acknowledges the support of Jane Addams, Sophonisba Breckinridge and Julia Lathrop, the most prominent members of the Hull House settlement, but also twenty-three other women and three men who collaborated on the project. All the chapters in the book are written by women – the research was almost entirely women's work.

THE CHICAGO HOUSING MARKET

The studies were concerned with low-income groups, mostly recent immigrants, and focused on inner-city tenement areas, with occasional references to peripheral low-income areas. The analyses covered tenancy, family composition, race and ethnicity, with additional material on the housing crisis during the Depression in the early 1930s. Although the empirical evidence was specific to Chicago, some of the theoretical arguments had considerable generality, notably in relation to the structuring of the property market, landlordism, racism, and neighbourhood change. It is these aspects of the study which I will consider here because they have particular resonance for students of the capitalist city. Interestingly, some of the book's conclusions are close to those made by DuBois, although I think that the analyses in *The Tenements of Chicago* are generally more penetrating.

Rent was the central issue in the Chicago studies, both as an indicator of the operation of the housing market and as a factor constraining the housing opportunities of different racial and ethnic groups. There is no reference to foundational theory, particularly Marxism, but explanations of variations in rents could have been derived from a structuralist reading of the capitalist city. The text is clearly quite different from those analyses which occupied the centre ground until the early 1970s, such as Burgess's organic representation of the urban housing market, neo-classical urban rent theories, like that of William Alonso,[36] or empirical studies based on neo-classical theory, such as Maurice Yeates's study of Chicago land values.[37]

The direction of the argument is clear in a chapter on tenement rents by Helen Rankin Jeter. She observes that an analysis of supply and demand relationships does not adequately explain observed increases in rents and goes on to suggest that:

> Another factor may have been present . . . i.e. increased land values due to anticipated future use of the land for commercial or industrial uses. Since assessed valuation and actual tax rates are based to a certain extent upon this anticipated use, the owner and the landlord may be justified in asking higher rents; but higher rents in turn justify increased capitalization and consequently higher valuation of the building, which again justifies higher taxes. This circular effect of anticipated land value, taxation, income and valuation of the building may conceivably go on *without regard to the continuing decay of the building or the economic necessities of the renters.*
>
> (p. 291)

Although Jeter does not acknowledge Marx, and would probably have rejected the label 'Marxist', it is evident that she recognized that rents could not be understood in terms of a perfect market but, rather, that rent in the capitalist property market contained a contradiction between exchange value and use value. Her analysis could be compared with David Harvey's early Marxist argument that

> the evolution of urban land-use patterns can be understood only in terms of the general

processes whereby society is pushed down some path (it knows not how) towards a pattern of social needs and human relationships (which are neither comprehended nor desired) by the blind forces of an evolving market system. The evolution of urban form is an integral part of this general process and rent, as a measure of the interpenetration of use values and exchange values, contributes notably to the unfolding of this process.[38]

Jeter's awareness is similarly evident in Abbott's account of landlordism. She recognized that in the poorer districts of Chicago the decision to buy, usually with a mortgage, was encouraged by a general rise in rents. Abbott demonstrates that the landlord in the tenement districts was typically of a similar economic status to the tenants and that profits were small:

The owners are only landlords renting out the apartments in the tenement building in which they have invested with the hope that the apartments can be rented for enough to pay the interest on the mortgage and something towards its redemption. This form of home ownership is far from being a sign of prosperity. . . the home owning family not infrequently lives in the smallest and poorest apartment in the building since this is the apartment that will be rented with difficulty, while the more desirable apartments are used to bring in the largest possible income and help pay off what is due on the mortgage.

(pp. 377–8)

Again, we might compare Harvey:

Landlords in the inner city housing market are not making huge profits. In fact, the evidence suggests that they are making less than they would elsewhere in the housing market . . . Some are unethical, of course, but good, rational, ethical landlord behaviour provides a relatively low rate of return.[39]

Edith Abbott recognized also that landlordism had a cultural and social basis. Reasons for purchase, and renting part of the property, included feelings of citizenship which derived from owning property; the difficulty of finding rented property with a large family; the insecurity of the rental sector; and the need to keep extended family and kin-groups together. In other words, landlordism for recent immigrants was related to the need to find security in an alien environment. Abbott's argument appears more convincing because it is supported by extracts from interviews and quite intimate observations. To give two examples:

Another widow was found sitting on the front step with the youngest of her thirteen children, weeping because she had been ordered to move the following week and she thought no 'boss' would rent to her because of her 'thirteen kids'

(p. 382)

An immigrant Italian workingman with six children who had been making payments on his house over a period of seven years, and who had just bought his first entirely new suit of clothes since undertaking the purchase of the home, said he bought the home because he had a 'big, big family' and he did not want any landlord to 'boss the children when they're loud'.

(p. 383)

These comments on the urban property market certainly suggest a different perspective from that of Park and Burgess. In particular, there is an understanding of the problems experienced by the immigrant working class, and the engagement of the women with Italian, Hungarian and other immigrants informs their analysis. This has been missed in histories of social geography. Thus, Jackson and Smith,[40] in one of the better surveys of urban social theory, acknowledge the ethnographies of the Chicago sociologists and Firey's study of sentiment and symbolism in Boston,[41] but make no mention of Addams or Abbott. However, they do stress the importance of the pragmatic tradition of James and Dewey of which these women became a part as a result of their association at Hull House.

The analysis of racism in *The Tenements of Chicago* is another strength of the study. Here, there is a marked contrast with Park's views on race. Abbott's analysis is not original,[42] but she makes her point strongly, linking her observations with arguments for social justice. Thus:

The color line, as it appears in the housing problem of Chicago and every other northern city, is too important to be overlooked . . . In the face of continued manifestations of race prejudice, the Negro has come to acquiesce silently as various civil rights are withheld from him . . . He rarely protests, for example, when he is refused entertainment at a restaurant or a hotel, or when he is virtually excluded from places of public entertainment, public places of recreation, and even the better shopping centers. There are three points, however, on which he cannot yield, even to keep the peace. He must claim a fair chance to find work for a living wage, good schools for his children, and a decent home for his family in a respectable neighborhood and at a reasonable rental.

(p. 118)

In many respects, she echoes DuBois, whom she does not acknowledge, particularly in the observation that racism contributes to the creation of housing sub-markets. The following quotation, for example, is almost identical to DuBois's argument on the creation of the black housing market in Philadelphia:

In any study of . . . housing conditions in Chicago, *the problem of the Negro will be found to be quite different from that of the immigrant groups.* With the Negro, the housing problem has long been an acute problem *not only* among the poor, as in the case of the Polish, the Jewish or the Italian immigrant but also among the well-to-do of the same racial group . . . The prejudice among the white people of having Negroes living on what they regard jealously as their residence streets and their unwillingness to have Negro children attending schools with white children [confine] the opportunities for residence open to Negroes *in all positions in life* to relatively small and well-defined areas. Consequently, the demand for houses and apartments within these areas is comparatively steady and, since the landlord is reasonably certain that the house or apartment can be filled at any time, as long as it is in any way tenantable, he takes advantage of his opportunities to raise rents and postpone repairs.

(p. 125)

Elsewhere, Helen Jeter makes a similar point. Like DuBois, she and Abbott recognize racism as a source of oppression which can be treated as a dimension separate from class. Racism cuts across class lines and, in Chicago in the 1920s, affected residential choice in a way that (white) ethnicity did not, creating a separate black housing market. This construction of the problem should be compared with Park's sanguine view of neighbourhood succession and the filtering of the urban, black population through the housing stock, referred to in the last chapter.

One further comment by Chicago sociologists serves to emphasize the distance between 'the Chicago School' and Edith Abbott and her associates. Burgess and Bogue,[43] discussing the inner city, maintained that 'the same community conditions that give rise to tuberculosis give rise to juvenile delinquency – non-white population, immigrant population, bad housing. All the factors of community deterioration that lead to one also lead to the other.' This is a typically 'neutral', non-problematic formulation, reflecting the primacy of the ecological model in informing the sociologists' view. There is no recognition of the importance of power relations or racism.

A further distinctive characteristic of Abbott's analysis of housing problems is that it was linked explicitly to practice. She recommended state intervention in the housing market as the only way of providing adequate accommodation for the urban population. Thus, she argues that

Chicago is not unlike New York in the vast extent of the slum areas. Public housing – and public housing on a very large scale – will be necessary to meet the situation. More than that, this public housing policy must be a national policy, with federal funds available for a large-scale housing program.

(p. 494)

Abbott then concludes with a more general argument for socialism:

The final answer, if and when it comes, to the ever present housing question must come from the economic side. The unskilled workers even in periods of prosperity do not have the wages to pay for decent houses. The employer must pay higher wages, or very wide areas must be cleared and very great numbers of new houses must be furnished out of taxes. There is no other way.

(pp. 494–495)

THE URBAN EXPERIENCE

Commentaries on urban social geography, including that by Jackson and Smith, use the term 'Chicago ethnography' to denote a specific method of enquiry associated with the School of Sociology. These writers, and others, like Ley,[44] have attempted to recover this method as part of a broader programme of developing a humanistic geography, but they have all neglected the sensitive descriptions of urban cultures which were produced by the Chicago women. A sympathy for and engagement with the oppressed was the central purpose of Hull House, as it was conceived by Jane Addams and Florence Kelly, and Edith Abbott's writing reflects this purpose. She was particularly skilled at depicting the problems faced by recent immigrants to the city, people from peasant backgrounds who found the alien environment and the individualism and materialism of Chicago difficult to come to terms with.

The following passage, for example, conveys the tensions and ambiguities that accompanied urban living in a Croatian community:

In material ways, the Croatians realize that they are far better off than they were in Europe; they have more money and can have better food and clothing and greater advantages than would ever have been possible there, but before the war many of them were homesick and wanted to go back . . . One of the old Croatian group said, 'When two people meet on the street, here they say "Got a good job? – How much do you get? Who's the foreman?" In Europe, if the same men met, they would have said, "What are you doing on Sunday?"' Here, they do not know what to do on a Sunday – there is no form of recreation that appeals to them. . . Here, they do not feel free in the parks, they do not know how to reach the country.

(p. 115)

Although Abbott documented urban change in a rather matter of fact way, she also had a feel for the social impact of restructuring. One vignette describes the eviction of Italian families by a railroad company. It conveys vividly the dislocation and stress which could be associated with the process of neighbourhood change:

The tenement was solidly Italian and when the notice was given to the Italian families to vacate the house, the order was received as a calamitous blow. . . How and where could they move? They complained that they had 'nowhere to go' and it was true that the housing accommodation in this particular area was limited because of the continuous encroachment and the constantly increasing demands of business. Most of the families had no contact with the other Italian colonies in other parts of the city. The evicted families finally refused to move, and the railroad authorities gave orders that their household goods were to be carried out and put on the sidewalk. This was actually done and the confusion of the noisy street, with women and children crying and sympathetic neighbours trying to offer consolation, was a scene of great disorder. . . The worst feature of the situation was that there were two hundred and fifty children living in this block . . . It was possible to find tenements for the evicted families in the West End Italian colony but they looked upon this journey one mile west and across the river as a journey into a far country.

(p. 112)

Such commentaries did not obscure the larger picture. From the surveys we get a fairly full picture of the spatial structure of the city, which is coupled with analyses of the economy and social relations that have considerable generality. However, the observations extend to such things as the details of furnishings in tenement rooms and accounts of scavenging on refuse tips, an aspect of the informal economy which does not feature in most accounts of the developing metropolis in the industrialized world. Ironically, Robert Park emphasized the value of such accounts of the experience of groups and individuals in the urban environment, and Bulmer stresses that Park was concerned to capture the world-views of people different from himself.[45] In fact, in one paper on the role of the organic analogy in understanding social systems,[46] he accorded priority to what he termed 'human documents':

It is because human actions must be interpreted in order to make them intelligible that documents – human documents – are more important for the study of human nature than statistics or formal facts. The documents are valuable, therefore, not merely because they describe events but because they throw light upon motives; that is to say, upon the subjective aspects of events and acts in which human nature manifests itself. Not merely events, but institutions as well become intelligible when we know their histories, and particularly when we know the individual experiences of men and women in which they had their origin and on which they finally rest.

This comment indicates the breadth of Park's vision – he could not be categorized simply as a 'scientific' sociologist – but it also makes it more surprising that Edith Abbott and her associates, who had been producing human documents since 1908, were not acknowledged by Park or in later histories of urban studies.

CONCLUSION

It is undeniable that the women working in the School of Social Service Administration at the University of Chicago, and earlier at Hull House, made significant contributions to an understanding of socio-spatial relations in cities. Their stated goals were, in fact, similar to those of the Chicago sociologists, that is, they were intent on developing a scientific study of urban society. Thus, Abbott, in 1931, asserted that 'if social research is to go on, it can only develop scientifically with the help of well-trained social workers', and in the same spirit, Breckinridge wrote in 1936 that

> an understanding of the principle of research and an attitude of mind recognizing the degree to which all social work instruction and social work practice is near what one might call 'the margin of the unknown' are essential for any adequate preparation [for social work].[47]

Referring to the role of settlements in research on social groups in Chicago, Diner maintained that they provided a great stimulus to the development of empirical social science.[48]

The general similarity in the approaches of the women and the male sociologists was denied by Park, however. The rather limited view of science that comes across in Park's writing could be interpreted as an attempt to erect a boundary between what he appeared to see as masculine and feminine spheres of knowledge. Analytical, scientific sociology is opposed to caring, practical social work, with the former ranked above social work research in a hierarchy of knowledges. I have suggested, however, that Park's dismissive attitude towards the research done in the School of Social Service Administration is not only an issue of the signing of knowledge. The radical positions taken by some of the Hull House residents and researchers, manifest in campaigns for women's rights, the anti-war movement, internationalism and support for immigrants, as well as their socialist programmes for housing, marked them as dangerous and subversive in conservative, nationalistic 1920s America. The political commitment of the women should not be seen as something distinct from their academic concerns. Addams, Abbott and Breckinridge recognized that theory and (political) practice were inextricably linked, whereas Robert Park, in the name of science, argued that theory and practice should be divorced. Involvement and detachment did distinguish the social work researchers and the sociologists, respectively.

It is unlikely that the cases discussed in this and the previous chapter are unique. Kivisto describes a similar case concerning the invisibility of Horace Kallen, a contemporary of Park who questioned the latter's assimilationist view of race relations and proposed a pluralist model.[49] Kivisto claims that Kallen was ignored because, among other reasons, he was not a part of the emerging scientific community; his writings reflected an obvious partisanship, at odds with calls for greater objectivity, and, as a Jew, he encountered anti-Semitic prejudice that served to marginalize his thought. As with the women, a devaluation of knowledge on the grounds that it was not 'scientific' was coupled with denigration on the grounds of social difference.

It might be argued that the professional success of Edith Abbott, her sister Grace, as head of the Children's Bureau in Washington, or Sophonisba Breckinridge, professor in the School of Social Service Administration, hardly suggests suppression or marginalization. Their academic work did not enter the mainstream of social science, however, whereas the contributions of Park and Burgess have canonical status because their theories contributed to positive and conservative sociologies and geographies which were dominant until the 1970s. The women's critical, left stance was clearly at variance with social science paradigms which prevailed until Marxist and other critical perspectives gained a foothold. Successful in a professional field deemed suitable for women, they were marginalized in a field dominated by rather conservative men.

The continuing appeal of Chicago sociology and the continuing failure, within human geography, to notice the research done by the School of Social Service Administration is interesting. What, then, are the resonances which Chicago sociology has had for modern geography and which the research done by the women has apparently lacked?

In geography, particularly British geography, there was no tradition of critical, politically informed work on the city before the 1970s. There was a landscape/fieldwork tradition, but this was primarily a rural pursuit only occasionally applied to cities in cosy morphological studies. When spatial science began to take shape in the late 1950s, it was this morphological tradition which was rejected in favour of an abstract, universalizing science. Spatial science was able to draw on neo-classical economics, but it was particularly short on spatial theories with universal claims. It is, therefore, unsurprising that Christaller's central place theory and Burgess's model of the city were revived because they were among the few placeless geometries which could be coupled with neo-classical economic theory. Spatial science secured a place for Chicago sociology, albeit in a rather thin and undernourished state.

It was the Chicago School of Sociology which was again seen as foundational for humanistic geography, but this time it was Chicago ethnography and John Dewey's pragmatic philosophy which inspired. Otherwise, critical human geography looked to Marx before opening up to more diverse philosophical argument. Thus, over the past forty years or so, totalizing theory has had a greater influence on the subject than the kind of localized knowledge produced by the School of Social Service Administration. Now, a human geography informed by feminist and post-modern theory should be more receptive to other voices from the past, like the Chicago women.

NOTES

1. This chapter is an expansion of two earlier papers: 'Invisible women? The contribution of the Chicago School of Social Service Administration to urban analysis', *Environment and Planning A*, 22, 1990, 733–745, and 'Gender, science, politics and geographies of the city', *Gender, Place and Culture*, 1995 forthcoming.

2. Peter Haggett, *Locational Analysis in Human Geography*, Edward Arnold, London, 1965.

3. Leo Schnore, 'On the spatial structure of cities in the two Americas', in Philip Hauser and Leo Schnore (eds), *The Study of Urbanization*, Wiley, New York, 1965, pp. 347–398.

4. Donna Haraway, 'Situated knowledges: the science question in feminism and the privilege of partial perspective', *Feminist Review*, 14 (3), 1988, 575–599.

5. Evelyn Fox Keller, 'Feminism and science', in E. Abel and E. Abel (eds), *The Signs Reader: Women, gender and scholarship*, Chicago University Press, Chicago, 1983, pp. 109–122.

6. Evelyn Fox Keller, 'Gender and science', *Psychoanalysis and Contemporary Thought*, 1, 1978, 409–433.

7. Keller, op. cit., 1983, pp. 115–116.

8. Iris Young, *Justice and the Politics of Difference*, Princeton University Press, Princeton, N.J., 1990.

9. Jo Benjamin, 'Bonds of love: rational violence and erotic domination', *Feminist Review*, 6 (1), 1980, 144–174.

10. For a discussion of the universal and the particular as they were debated by ancient Greek philosophers see Paul Feyerabend, *Farewell to Reason*, Verso, London, 1987, chapters 2 and 6.

11. Elizabeth Potter, 'Modelling the gender politics of science', *Hypatia*, 3 (1), 1988, 19–33.

12. Hermeticism derives from Hermes Trismegistus whose writing was in fact the collective work of a group of mediaeval scholars. This group subscribed to the natural magic tradition.

13. Edith Abbott (ed.), *The Tenements of Chicago, 1908–1935*, Chicago University Press, Chicago, 1936.

14. *Residents of Hull House*, Chicago, 1895, pp. 207–208.

15. ibid.

16. Christopher Lasch, *The New Radicalism in America, 1889–1963: The intellectual as social type*, Chatto and Windus, London, 1966, p. xv.

17. Allen Davis, *American Heroine: The life and legend of Jane Addams*, Oxford University Press, New York, 1973, p. 97.

18. Jane Addams, *Twenty Years at Hull House*, Macmillan, New York, 1911, pp. 402–403.

19. Jane Addams, *The Second Twenty Years at Hull House*, Macmillan, New York, 1930, p. 80.

20. World's Fairs in the late nineteenth and early twentieth centuries celebrated the achievements of capitalism and imperialism. They were, inevitably, male-dominated. The Women's World's Fairs in Chicago challenged the basis of the established fairs.

21. Mary Jo Deegan, *Jane Addams and the Men of the Chicago School, 1892–1918*, Transaction Books, New Brunswick, N.J., 1988, p. 310.

22. Mary Jo Deegan, 'Early women sociologists and the American Sociological Society: the patterns of exclusion and participation', *American Sociologist*, 16, 1981, 14–24.

23. Addams, 1930, op. cit., pp. 140–141.

24. Davis, op. cit., p. 252.

25. ibid., p. 254.

26. Steven Queen, 'Seventy-five years of American sociology in relation to social work', *American Sociologist*, 16 (1), 1981, 34–37.

27. Martin Bulmer, *The Chicago School of Sociology*, Chicago University Press, Chicago, 1984, p. 68.

28. Ernest Burgess, 'The contribution of sociology to family social work', *Family*, 8, 1927, 191–193.

29. Robert Faris, for example, noted that

> The tradition is strongly developed in the United States that sociology should be developed in much the same spirit as a 'pure science'. This does not mean that it must be a useless science or that it must have no known connection with practical matters but the research itself should be done in a mood of detachment, with the conclusions following from the study and not necessarily resulting in an immediate reform or welfare program.

See Steven Diner, 'Scholarship in the quest for social welfare: a fifty-year history of the *Social Service Review*', *Social Service Review*, 1977, 1–66.

30. ibid.

31. Ernest Burgess and Donald Bogue, *Contributions to Urban Sociology*, Chicago University Press, Chicago, 1964.

32. Nicholas Entrikin, 'Robert Park's human ecology and human geography', *Annals, Association of American Geographers*, 70 (1), 1980, 46.

33. ibid., 43–58.

34. Deegan, op. cit., 1988, p. 310.

35. Lela Costin, *Two Sisters for Social Justice: A biography of Grace and Edith Abbott*, University of Illinois Press, Urbana, Ill., 1983, p. 21. Costin includes a full listing of Edith Abbott's published work.

36. William Alonso, 'A theory of the urban land market', *Papers and Proceedings of the Regional Science Association*, 6, 1960, 149–157.

37. Maurice Yeates, 'Some factors affecting the spatial distribution of Chicago land values,

1910–1960', *Economic Geography*, 41 (1), 1965, 57–70.

38. David Harvey, *Social Justice and the City*, Edward Arnold, London, 1973. This seems an appropriate text to use for comparison with Jeter's writing because it was greeted as an innovative theoretical contribution to urban studies in the 1970s. To my knowledge, Harvey has never referred to the Chicago women.

39. ibid., p. 140.

40. Peter Jackson and Susan Smith, *Exploring Social Geography*, George Allen and Unwin, London, 1984.

41. W. Firey, 'Sentiment and symbolism as ecological variables', *American Sociological Review*, 10, 1945, 140–148.

42. Abbott's and Jeter's perspectives on race echoed earlier views expressed by settlement house workers and their associates. Louise de Koven Bowen, who donated over one million dollars to Hull House, wrote *The Colored People of Chicago* in 1913. In this report, she identified the problem of discrimination in the housing market and, more generally, that

> the negro's position is more difficult than that [of foreigners] for he is subjected to racial discrimination and, while no limitations are imposed upon the children of the immigrant, this is unfortunately not true of the children of the negro.

Jane Addams, who recognized the destructive effects of slavery on American black culture and likewise acknowledged the strengths and values of African cultures, worked on behalf of blacks through the NAACP and the Chicago Urban League. Sophonsiba Breckinridge wrote a short article on discrimination against blacks in the housing market in 1913. See Diner, op. cit., 1977.

43. Burgess and Bogue, op. cit., p. 9.

44. David Ley, 'Social geography and the taken-for-granted world', *Transactions, Institute of British Geographers*, NS, 2, 1977, 498–512.

45. Bulmer, op. cit., p. 77. According to Kivisto, however, Park's vision was limited because of his assimilationist perspective on ethnic minorities.

> I should argue that, insofar as assimilation became the dominant theoretical construct, interpretative, hermeneutic approaches to ethnicity were not pursued to any significant extent – and this despite the fact that both Park and Thomas encouraged modes of interpretative sociology. Not only did this result in the devaluation of investigations into the immigrant life-world, but it failed to provide an account of human agency that would treat immigrants as authors of their own lives.
>
> (Peter Kivisto, 'The transplanted then and now: the reorientation of immigration studies from the Chicago School to the new social history', *Ethnic and Racial Studies*, 13 (4), 1990, 455–481)

46. Robert Park, 'Human nature and collective behavior', *American Journal of Sociology*, 1926, 740.

47. Diner, 1977, op. cit., p. 11.

48. Steven Diner, *A City and Its Universities: Public policy in Chicago, 1892–1919*, University of North Carolina Press, Chapel Hill, 1980, p. 123.

49. Kivisto, op. cit.

CONCLUSION

> But the city in its corruption refused to submit to the dominion of the cartographers, changing shape at will and without warning.
>
> (Salman Rushdie, *The Satanic Verses*, p. 327)

One assertion which I make at various junctures is that people's relationships with others are too often conditioned by fear and that fear, anxiety, nervousness also affect attitudes to knowledge. New ideas or subversive ideas can be as threatening as images of alien others. This reaction to certain kinds of difference is bound up with questions of power. A fear of mixing unlike things often signifies a reluctance to give ground and relinquish power. In all kinds of political, social and socio-spatial relationships, boundaries then assume considerable significance because they are simultaneously zones of uncertainty and security. Policing boundaries is one way of reducing fear. Boundary crossing is discouraged by appeals to loyalty – to the state, to the community, to the (academic) subject, and so on. The maintenance of secure borders is not always easy, however, because groups who are fearful of mixing and heterogeneity may lack the power to control entry to their space.

It might appear that this problem, the subject of this book, has been largely eclipsed by forces which have breached old boundaries and created a world of fractured, hybridized and fused identities. For example, the end of the cold war has rendered a particularly powerful rhetoric which supported a boundary between 'good' and 'evil' redundant. Migrations of peoples and cultures have given the South a much more influential presence in the North than in the past and not just in established cosmopolitan centres like London, Paris or New York. In the academy, post-modern texts have blurred previous subject identities.

I doubt, however, whether these cultural, political and social transformations have really made people less fearful, less concerned about keeping a distance from others, less exclusionary in their behaviour. The world political map in 1994 is replete with new, strong boundaries which are designed to secure cultural homogeneity, and, at the local level, hostility towards outsider groups like New Age Travellers in England and Wales and ethnic minorities in much of Europe is no less acute than it was before 'the passing of the modern world'. The desire for a purified identity, which requires the distant presence of a bad object, a discrepant other, seems to be unaffected by cross-currents of culture which are characteristic of recent global change.

The social sciences, and human geography in particular, might now be better equipped to challenge xenophobia, racism and other exclusionary tendencies because of a greater intellectual awareness of difference. The feminist literature on difference, anthropological texts by James Clifford, Constance Perin and others, work by black authors, and so on have been acknowledged and have contributed to geographical writing which shows a growing sensitivity to other voices. Derek Gregory,[1] for example, recognizes that moving from (multicultural) Britain to multicultural Canada and becoming aware of the tensions of a multicultural society (for the first time?) made him conscious of his 'own otherness, troublingly and imperfectly'. However, there is still a distance between authors (mostly male) and their subjects. Post-modern discourse does not bring the academic writer closer to the 'other' if there is no real engagement. Engagement with texts does not remove the need for engagement with people. In other words, I see the question of making human geography radical and emancipatory partly as a question of getting close to other people, listening to them, making way for them.

The methods of enquiry which this demands – participant observation, group work, grass-roots activism or learning through advocacy – are not practised by most geographers. They are, admittedly, difficult to reconcile with the demands of institutions operating according to market principles. Understanding and high productivity rarely go together, and long-term involvement with another culture might seem a not very productive use of time (although social anthropologists seem to manage).

The problem that the subject has in coming to grips with difference in a useful and meaningful way is related to the methods used by geographers and to anxieties and inhibitions which these methods do nothing to dispel. Admittedly, a preoccupation with cartography is much less evident than it was. The subject is not disciplining the world through maps, imposing order and denying ambiguity to the extent that it did when the rules were written according to the laws of spatial science. However, there is still an often excessive distance between the observer and the observed which means that the categorical schemes of the former are unchallenged by the experience of others and by other world-views. The voyeuristic impulse, which Gillian Rose represents as the 'masculine gaze', reifying and dehumanizing 'the other',[2] has a particular expression in cartography and other forms of distant geographical description. Mapping provided security: the cartographer was in control and would not be disturbed by other people's maps or alternative world-views. But there is a danger that the current pre-occupation with texts, and particularly with what Judith Still calls 'author-figures' (Marx, Foucault, Derrida, etc.),[3] will have the same damaging effect. Apart from perpetuating an elitist and hierarchical view of knowledge, debates about author-figures are also safe. There is not the same risk of failure that accompanies an encounter with a person whose world-view is different from your own. If such encounters are to lead to productive dialogue, it is necessary to open up and to accept that the experience may be disturbing. The disciplinary context within which most geographers work seems to me inhibiting, making the subject itself exclusionary. Without wishing to be too prescriptive, I would suggest that if geography is to represent difference authentically and to challenge exclusionary tendencies, practitioners need to transgress disciplinary and personal boundaries and to come much closer to the people whose problems provide the primary justification for the existence of the subject.

What I would advocate is that geographers go out into the world (I recognize that some are already there), not on an imperialist and colonialist mission, but in order to experience the life-worlds of other people. This is hardly a novel suggestion,[4] but I do not think that the implications for theory and practice have been adequately discussed. Primarily, experiencing other people's worlds requires that fear of others is overcome. At this point, psychoanalytical theory has considerable value because it can help us to understand better not only the representation of others but also our own feelings about the abject, our own insecurities about difference which affect our academic practice. Apart from the intellectual attraction of psychoanalytical theory in articulating the connection between the self and the social and material world, I have found some ideas coming particularly from object relations theorists compelling because they have helped me to appreciate my own fears and inhibitions associated with places and other people. They also connect readily with theoretical arguments about the symbolic role of boundaries, which come primarily from social anthropology. The 'psychoanalytical turn', rather than being a fad as this term suggests, for me provides a vocabulary and cues for observation and analysis which are helpful in getting to grips with difference. Understanding the experience of others and their relationship to place involves positioning ourselves in the world. Listening to and talking with people is one necessary part of this endeavour. Reflecting on the experience in such a way that we recognize our own part in the dialogue is another.

NOTES

1. Derek Gregory, *Geographical Imaginations*, Basil Blackwell, Oxford, 1994, p. ix.

2. Gillian Rose, *Feminism and Geography*, Polity Press, Cambridge, 1993. To claim that 'the gaze' is a masculine condition essentializes. Racism, for example, could also be identified as a problem of distancing and mis-representation – a consequence of gazing from afar on subject colonial peoples. The representation of women in a masculinized geography is one of a number of related problems.

3. Judith Still, 'What Foucault fails to acknowledge . . .: feminists and *The History of Sexuality*', *History of the Human Sciences*, 7 (2), 1994, 150–157.

4. I am thinking here particularly of Bill Bunge's geographical expeditions, the 1970s Byker conference on Tyneside, which was designed to bring geography to the people, and David Ley's writing during the same period, for example, his 'Social geography and the taken-for-granted world', *Transactions, Institute of British Geographers*, NS, 2, 1977, 498–512.

BIBLIOGRAPHY

Abbott, E. (ed.), *The Tenements of Chicago, 1908–1935*, Chicago University Press, Chicago, 1936.

Addams, J., *Twenty Years at Hull House*, Macmillan, New York, 1911.

Addams, J., *The Second Twenty Years at Hull House*, Macmillan, New York, 1930.

Adorno, T. *et. al.*, *The Authoritarian Personality*, Norton, New York, 1982.

Alaszewski, A., *Institutional Care and the Mentally Handicapped: The mental handicap hospital*, Croom Helm, London, 1986.

Alonso, W., 'A theory of the urban land market', *Papers and Proceedings of the Regional Science Association*, 6, 1960, 149–157.

Altman, I., *The Environment and Social Behavior*, Brooks Cole, Monterey, Calif., 1975.

Ament, D. and van der Knapp, G., 'A two-stage model for the analysis of intra-urban mobility processes', *Environment and Planning A*, 17, 1985, 1201–1216.

Anderson, K. and Gale, F. (eds), *Inventing Places: Studies in cultural geography*, Longman Cheshire, Melbourne, 1992.

Anglemont, P. d', 'La Ville des chiffonniers', *Paris Anecdote*, 1854, 218.

Ardener, S. (ed.), *Perceiving Women*, Dent, London, 1977.

Atkinson, P., *Language, Structure and Reproduction: An introduction to the sociology of Basil Bernstein*, Methuen, Andover, 1985.

Babcock, B. (ed.), *The Reversible World: Symbolic inversion in art and society*, Cornell University Press, Ithaca, 1978.

Bachelard, G., *La Poétique de l'espace*, Presses Universitaires de France, Paris, 1981.

Badcock, B., *Unfairly Structured Cities*, Basil Blackwell, Oxford, 1984.

Baldasarre, M., *Trouble in Paradise: The suburban transformation of America*, Columbia University Press, New York, 1986.

Barnes, B., *Scientific Knowledge and Sociological Theory*, Routledge and Kegan Paul, London, 1974.

Barnes, B., *T. S. Kuhn and Social Science*, Macmillan, London, 1982.

Barnes, T. and Duncan, J. (eds), *Writing Worlds: Discourse, text and metaphor in the representation of landscape*, Routledge, London, 1992.

Barrett, M., 'The concept of difference', *Feminist Review*, 26, 1987, 29–41.

Bauman, Z., 'Semiotics and the function of culture', in Julia Kristeva *et. al.* (eds) *Essays in Semiotics*, Mouton, The Hague, 1971, 279–295.

Becher, A., *Academic Tribes and Territories*, Open University Press, Milton Keynes, 1989.

Benjamin, J., 'Bonds of love: rational violence and erotic domination', *Feminist Review*, 6 (1), 1980, 144–174.

Bernstein, B. 'Open schools, open society?' *New Society*, 14 September 1967, 351–353.

Bernstein, B., 'On the classification and framing of educational knowledge' British Sociological Association Conference on Sociology of Education, 1970.

Bernstein, B., *Class, Codes and Control*, vol. 1, Paladin, St Albans, 1971.

Bhabha, H., *The Location of Culture*, Routledge, London, 1994.

Blacker, C., *Eugenics: Galton and after*, Duckworth, London, 1952.

Blackwell, J. and Janowitz, M., *Black Sociologists: Historical and contemporary perspectives*, Chicago University Press, Chicago, 1974.

Blonsky, M. (ed.), *On Signs*, Basil Blackwell, Oxford, 1985.

Booth, C., *Life and Labour of the London Poor*, 3rd Series, Religious Influences, Macmillan, London, 1902.

Brody, H., *The People's Land: Eskimos and whites in the eastern Arctic*, Penguin, Harmondsworth, 1975.

Brody, H., *Maps and Dreams: Indians and the British Columbia frontier*, Jill Norman and Hobhouse, London, 1981.

Bulmer, M., 'Charles S. Johnson, Robert E. Park, and the research methods of the Chicago Commission on Race Relations, 1919–1922: an early experiment in applied social research', *Ethnic and Racial Studies*, 4 (3), 1981, 289–306.

Bulmer, M., *The Chicago School of Sociology*, Chicago University Press, Chicago, 1984.

Bunge, W., *Fitzgerald: Geography of a revolution*, Schenkman, Cambridge, Mass., 1971.

Burgess, E., 'The contribution of sociology to family social work', *Family*, 8, 1927, 191–193.

Burgess, E and Bogue, D., *Contributions to Urban Sociology*, Chicago University Press, Chicago, 1964.

Burkitt, I., 'The shifting concept of the self', *History of the Human Sciences*, 7 (2), 1994, 7–28.

Capra, F., *The Turning Point: Science, society and the rising culture*, Bantam Books, London, 1987.

Certeau, M. de, 'Practices of space', in M. Blonsky (ed.), *On Signs*, Basil Blackwell, Oxford, 1985, 122–145.

Chambers, I., *Migrancy, Culture and Identity*, Routledge, London, 1994.

Christie, R. and Jahoda, M. (eds), *Studies in the Scope and Method of the Authoritarian Personality*, Free Press, Glencoe, 1954.

Clifford, J. and Marcus, G. (eds), *Writing Culture: The politics and poetics of ethnography*, University of California Press, Berkeley, 1986.

Cloke, P., Philo, C. and Sadler, D., *Approaching Human Geography*, Paul Chapman, London, 1991.

Cohen, A. (ed.), *Symbolizing Boundaries*, Manchester University Press, Manchester, 1986.

Cohen, S., *Folk Devils and Moral Panics*, MacGibbon and Kee, London, 1972.

Cohen, S., *Visions of Social Control*, Polity Press, Cambridge, 1985.

Cohen, S. and Taylor, L., *Escape Attempts*, Allen Lane, London, 1976.

Cooper, M., 'Making changes', in T. Putnam and C. Newton (eds) (1990) *Household Choices*, Futures Publications, London, 1990, 37–42.

Corbin, A., *The Fragrant and the Foul: Odor and the French social imagination*, Harvard University Press, Cambridge, Mass., 1986.

Costin, L., *Two Sisters for Social Justice: A biography of Grace and Edith Abbott*, University of Illinois Press, Urbana, Ill., 1993.

Csikszentmihalyi, M. and Rochberg-Halton, E., *The Meaning of Things: Domestic symbols and the self*, Cambridge University Press, Cambridge, 1981.

Davis, A., *American Heroine: The life and legend of Jane Addams*, Oxford University Press, New York, 1973.

Davis, J. C., *Fear, Myth and History: The Ranters and the historians*, Cambridge University Press, Cambridge, 1986.

Davis, M., *City of Quartz*, Verso Press, London, 1990.

Davis, N. and Anderson, B., *Social Control: The production of deviance in the modern state*, Irvington, N.Y., 1983.

Davis, N. Z., 'Women on top: symbolic sexual inversion and political disorder in early modern Europe', in B. Babcock (ed.), *The Reversible World: Symbolic inversion in art and society*, Cornell University Press, Ithaca, 1978, 147–190.

Deegan, M., 'Early women sociologists and the American Sociological Society: the patterns of exclusion and participation', *American Sociologist*, 16, 1981, 14–24.

Deegan, M., *Jane Addams and the Men of the Chicago School, 1892–1918*, Transaction Books, New Brunswick, N.J., 1988.

Diner, S., 'Chicago social workers and blacks in the Progressive era', *Social Service Review*, 44, 1970, 393–410.

Diner, S., 'Scholarship in the quest for social welfare: a fifty-year history of the *Social Service Review*, *Social Service Review*, 1977, 1–66.

Diner, S., *A City and its Universities: Public policy in Chicago, 1892–1919*, University of North Carolina Press, Chapel Hill, 1980.

Douglas, M., *Purity and Danger*, Routledge and Kegan Paul, London, 1966.

Douglas, M., *Natural Symbols*, Barrie and Jenkins, London, 1970.

Dovey, K., 'Homes and homelessness', in I. Altman and C. Werner (eds), *Home Environments*, Plenum Press, New York, 1985, 33–61.

Driver, F., 'Moral geographies: social science and the urban environment in mid-nineteenth century England', *Transactions, Institute of British Geographers*, NS, 13 (4), 1988, 275–287.

DuBois, W. E. B., *The Philadelphia Negro: A social study*, University of Pennsylvania Series in Political Economy and Public Law 14, Benjamin Blom, New York, 1967 (first published 1899).

Dyer, R., *The Matter of Images: Essays on representation*, Routledge, London, 1993.

Edwards, G. F., 'E. Franklin Frazier', in J. Blackwell and M. Janowitz (eds), *Black Sociologists: Historical and contemporary perspectives*, Chicago University Press, Chicago, 1974, 85–117.

Elias, N., *The Society of Individuals*, Basil Blackwell, Oxford, 1991.

Elliott, A., *Social Theory and Psychoanalysis in Transition*, Blackwell, Oxford, 1992.

Ellman, M., 'Eliot's abjection', in J. Fletcher and A. Benjamin (eds), *Abjection, Melancholia and Love: The work of Julia Kristeva*, Routledge, London, 1990, 178–200.

Entrikin, N., 'Robert Park's human ecology and human geography', *Annals, Association of American Geographers*, 70 (1), 1980, 43–58.

Erikson, E., *Childhood and Society*, Penguin, Harmondsworth, 1970.

Evans, D. and Herbert, D. (eds), *The Geography of Crime*, Routledge, London, 1989.

Eyles, J. and Lee, R., 'Human geography in explanation', *Transactions, Institute of British Geographers*, NS, 7 (1), 1982, 117–122.

Ferrero, G., *Criminal Man*, The Knickerbocker Press, New York, 1911.

Feyerabend, P., *Farewell to Reason*, Verso, London, 1987.

Feyerabend, P., *Against Method*, Verso Press, London, 1988.

Firey, W., 'Sentiment and symbolism as ecological variables', *American Sociological Review*, 10, 1945, 140–148.

Foucault, M, *Power/Knowledge*, Pantheon, New York, 1980.

Foucault, M., 'Of other spaces', *Diacritics*, 16 (1), 1986, 22–27.

Freud, S., *Civilization and Its Discontents*, Norton, New York (1961 edn).

Friedman, J., *The Monstrous Races in Medieval Art and Thought*, Harvard University Press, Cambridge, Mass., 1981.

Fyfe, G. and Law, J., *Picturing Power: Notes on the politics of visualization*, Sociological Review Monograph 35, Routledge, London, 1988.

Geertz, C., 'Blurred genres: the refiguration of social thought', *American Scholar*, 49, 1980, 165–178.

Geremek, B., *The Margins of Society in late-Mediaeval Paris*, Cambridge University Press, Cambridge, 1987.

Giddens, A., *Capitalism and Modern Social Theory*, Cambridge University Press, Cambridge, 1971.

Giddens, A., *Profiles and Critiques in Social Theory*, University of California Press, Berkeley, 1982.

Giddens, A. *The Constitution of Society*, Polity Press, Cambridge, 1984.

Gilman, S., *Difference and Pathology: Stereotypes of sexuality, race and madness*, Cornell University Press, Ithaca, N.Y., 1985.

Gilman, S., *Disease and Representation*, Cornell University Press, Ithaca, N.Y., 1988.

Gilman, S., *Sexuality: An illustrated history*, Wiley, Chichester, 1989.

Gilroy, P., 'The end of anti-racism', *New Community*, 17 (1), 1990, 71–83.

Gregory, D., *Geographical Imaginations*, Basil Blackwell, Oxford, 1994.

Gregory, D. and Urry, J. (eds), *Social Relations and Spatial Structures*, Macmillan, London, 1985.

Gregson, N., 'Structuration theory: some thoughts on the possibilities for empirical research', *Environment and Planning D: Society and Space*, 5, 1987, 73–91.

Gross, E., 'The body of signification' in J. Fletcher and A. Benjamin (eds), *Abjection, Melancholia and Love: The work of Julia Kristeva*, Routledge, London, 1990, 80–103.

Gross, E., 'Julia Kristeva', in E. Wright (ed.), *Feminism and Psychoanalysis: A critical dictionary*, Basil Blackwell, Oxford, 1992, 194–200.

Hacking, I., 'The archeology of Foucault', in D. Hoy (ed.), *Foucault: A critical reader*, Basil Blackwell, Oxford, 1986, 27–40.

Haggett, P., *Locational Analysis in Human Geography*, Edward Arnold, London, 1965.

Haggett, P. and Chorley, R., *Models in Geography*, Methuen, London, 1967.

Haraway, D., 'Situated knowledges: the science question in feminism and the privilege of partial perspective', *Feminist Review*, 14 (3), 1988, 575–599.

Hardy, D. and Ward, C., *Arcadia for All: The legacy of a makeshift landscape*, Mansell, London, 1984.

Harvey, D., *Social Justice and the City*, Edward Arnold, London, 1973.

Harvey, D., *The Urban Experience*, Basil Blackwell, Oxford, 1989.

Harvey, D., 'Postmodern morality plays', *Antipode*, 24, 1992, 300–326.

Hauser, P. and Schnore, L. (eds), *The Study of Urbanization*, Wiley, New York, 1965.

Hinnant, C. H., *Purity and Defilement in Gulliver's Travels*, Macmillan, Basingstoke, 1987.

Hirsch, F., *Social Limits to Growth*, Routledge, London, 1978.

Hoggart, K. and Kofman, E. (eds), *Politics, Geography and Social Stratification*, Croom Helm, London, 1986.

Hoggett, P., 'A place for experience: a psychoanalytic perspective on boundary, identity and culture', *Environment and Planning D: Society and Space*, 10, 1992, 345–356.

Holmes, S. J. and Robins, L. N., 'The influence of childhood disciplinary experience on the development of alcoholism and depression', *Journal of Child Psychology and Psychiatry*, 28 (3), 1987, 399–415.

hooks, b., *Black Looks: Race and representation*, Turnaround, London, 1992.

Hoy, D., 'Power, repression, progress: Foucault, Lukes and the Frankfurt School', in D. Hoy (ed.), *Foucault: A critical reader*, Basil Blackwell, Oxford, 1986, 123–147.

Husbands, C., 'East End racism, 1900–1980', *The London Journal*, 8, 1982, 3–26.

Ignatieff, M., *A Just Measure of Pain: The penitentiary in the Industrial Revolution, 1750–1850*, Macmillan, Basingstoke, 1978.

Illich, I., *H_2O and the Waters of Forgetfulness*, Dallas Institute of Humanities and Culture, Dallas, 1984.

Jackson, P., 'Street life: the politics of Carnival', *Environment and Planning D: Society and Space*, 6, 1988, 213–227.

Jackson, P., *Maps of Meaning: An introduction to cultural geography*, Unwin Hyman, London, 1989.

Jackson, P. and Smith, S., *Exploring Social Geography*, George Allen and Unwin, London, 1984.

Jackson, R., *Fantasy: The literature of subversion*, Methuen, London, 1981.

James, A., 'Learning to belong: the boundaries of adolescence', in A. Cohen (ed.), *Symbolizing Boundaries*, Manchester University Press, Manchester, 1986, 155–170.

Johnston, K., 'Dangerous knowledge: a case study in the social control of knowledge', *Australian and New Zealand Journal of Sociology*, 14 (2), 1978, 104–113.

Johnston, R. J., 'On the nature of explanation in human geography', *Transactions, Institute of British Geographers*, NS, 5 (4), 1980, 402–413.

Johnston, R. J., 'To the ends of the earth' in R. J. Johnston (ed.), *The Future of Geography*, Methuen, London, 1985, 326–338.

Johnston, R. J., *Philosophy and Human Geography*, Edward Arnold, London, 1986.

Kahane, C., 'Object relations theory', in E. Wright (ed.), *Feminism and Psychoanalysis: A critical dictionary*, Basil Blackwell, Oxford, 1992, 284–290.

Kearney, M., 'Borders and boundaries of the state and self at the end of empire', *Journal of Historical Sociology*, 4 (1), 1991, 52–74.

Keen, B., *The Aztec Image in Western Thought*, Rutgers University Press, New Brunswick, N.J., 1971.

Keller, E., 'Gender and science', *Psychoanalysis and Contemporary Thought*, 1, 1978, 409–433.

Keller, E., 'Feminism and science', in E. Abel and E. Abel (eds), *The Signs Reader: Women, gender and scholarship*, Chicago University Press, Chicago, 1983, 109–122.

Kennan, G., 'The sources of Soviet power', *Foreign Affairs*, 25 (4), 1947.

Kinsman, P., *Landscapes of National Non-Identity*, Working Paper 17, Department of Geography, University of Nottingham, 1993.

Kivisto, P., 'The transplanted then and now: the reorientation of immigration studies from the Chicago School to the new social history', *Ethnic and Racial Studies*, 13 (4), 1990, 455–481.

Klein, M., *Our Adult World and Its Roots in Infancy*, Tavistock Pamphlet 2, London, 1960.

Korosec-Serfaty, P., 'The home from attic to cellar', *Journal of Environmental Psychology*, 4 (4), 1984, 172–179.

Kristeva, J., *Powers of Horror*, Columbia University Press, New York, 1982.

Kristeva, J., *Strangers to Ourselves*, Columbia University Press, New York, 1991.

Kristeva, J. *et.al.* (eds), *Essays in Semiotics*, Mouton, The Hague, 1971.

Kropotkin, P., *Modern Science and Anarchism*, Simian, London, no date.

Kuhn, T., *The Structure of Scientific Revolutions*, 2nd edn, Chicago University Press, Chicago, 1970.

Kunzle, D., 'World Upside Down: the iconography of a European broadsheet type' in B. Babcock (ed.), *The Reversible World: Symbolic inversion in art and society*, Cornell University Press, Ithaca, 1978, 39–94.

Lal, B. B., *The Romance of Culture in an Urban Civilization*, Routledge, London, 1990.

Lasch, C., *The New Radicalism in America, 1889–1963: The intellectual as a social type*, Chatto and Windus, London, 1966.

Law, J. and Whittaker, J., 'On the art of representation: notes on the politics of visualisation', in G. Fyfe and J. Law (eds), *Picturing Power: Visual depictions and social relations*, Sociological Review Monograph 35, Routledge, London, 1988.

Lawrence, R. J., *Housing, Dwellings and Homes*, Wiley, Chichester, 1987.

Leach, E., *Culture and Communication*, Cambridge University Press, Cambridge, 1976.

Lévi-Strauss, C., *The Savage Mind*, Weidenfeld and Nicolson, London, 1966.

Levine, D., *The Flight from Ambiguity*, Chicago University Press, Chicago, 1985.

Ley, D., 'Social geography and the taken-for-granted world', *Transactions, Institute of British Geographers*, NS, 2, 1977, 498–512.

Liègeois, J-P., *Gypsies: An illustrated history*, Al Saqi Books, London, 1986.

Lowe, G., Foxcroft, D. and Sibley, D., *Adolescent Drinking and Family Life*, Harwood Academic, Chur, Switzerland, 1993.

Lowe, R. and Shaw, W., *Travellers: Voices of the New Age nomads*, Fourth Estate, London, 1993.

Lowman, J., 'The geography of social control: clarifying some themes', in D. Evans and D. Herbert (eds), *The Geography of Crime*, Routledge, London, 1989, 228–259.

Markus, T., *Buildings and Power: Freedom and control in the origin of modern building types*, Routledge, London, 1993.

Mason, P., *Deconstructing America: Representations of the other*, Routledge, London, 1990.

Matthews, F., *Robert E. Park and the Chicago School*, McGill–Queen's University Press, Montreal, 1977.

Mayall, D., *Gypsy-Travellers in Nineteenth Century Society*, Cambridge University Press, Cambridge, 1988.

Mead, G. H., *Mind, Self and Society*, Chicago University Press, Chicago, 1934.

Meadows, D., *Nattering in Paradise: A word from the suburbs*, Simon and Schuster, London, 1988.

Merquior, J., *Foucault*, Fontana Press, London, 1985.

Minuchin, S., *Families and Family Therapy*, Tavistock, London, 1974.

Mitchell, O. and Smith, N., 'Bringing in race', *Professional Geographer*, 42 (2), 1990, 232–234.

Moore, J., *W. E. B. DuBois*, Twayne Publishers, Boston, 1981.

Moos, A. I. and Dear, M. J., 'Structuration theory in urban analysis, 1: theoretical exegesis', *Environment and Planning A*, 18, 1986, 231–252.

Morrill R. L., 'The Negro ghetto: problems and alternatives', *Geographical Review*, 55, 1965, 339–361.

Mumford, L., *The City in History*, Secker and Warburg, London, 1961.

Murray, S., 'Fuzzy sets and abominations', *Man*, 18, 1983, 396–399.

Neisser, U., 'Five kinds of self-knowledge', *Philosophical Psychology*, 1 (1), 1988, 35–59.

Nesbit, M., *Atget's Seven Albums*, Yale University Press, New Haven, 1992.

Neumann, I., 'Russia as Central Europe's constituting other', *East European Politics and Societies*, spring 1994, 349–369.

Neusner, J., *The Idea of Purity in Ancient Judaism*, E. J. Brill, Leiden, 1973.

Nicholson, L. J., *Feminism/Postmodernism*, Routledge, London, 1990.

Nozick, R., *Anarchy, State and Utopia*, Basil Blackwell, Oxford, 1974.

Okely, J., 'Gypsy women: models in conflict', in S. Ardener (ed.), *Perceiving Women*, Dent, London, 1975, 55–86.

Olson, D., Sprenkle, D. and Russell, C., 'Circumplex model of families and family systems', *Family Process*, 18, 1979, 2–28.

Ó Tuathail, G., 'Critical geopolitics and development theory: intensifying the dialogue', *Transactions, Institute of British Geographers*, NS, 19, 1994, 228–238.

Ovenden, G., *Victorian Children*, Academy Editions, London, 1971.

Park, R., *The City*, Chicago University Press, Chicago, 1925.

Park, R., 'Human nature and collective behavior', *American Journal of Sociology*, 1926, 733–741.

Park, R. and Burgess, E., *An Introduction to the Science of Sociology*, Chicago University Press, Chicago, 1921.

Perin, C., *Everything in Its Place*, Princeton University Press, Princeton, 1977.

Perin, C., *Belonging in America*, University of Wisconsin Press, Madison, 1988.

Philo, C., *The Same and Other: On geographies, madness and outsiders*, Loughborough University, Department of Geography, Occasional Paper 11, 1987.

Pickles, J., 'Texts, hermeneutics and propaganda maps', in T. Barnes and J. Duncan (eds), *Writing Worlds: Discourse, text and metaphor in the representation of landscape*, Routledge, London, 1992, 193–200.

Pietz, W., 'The post-colonialism of cold war discourse', *Social Text*, 1988, 55–75.

Pile, S., 'Human agency and human geography re-visited: a critique of new models of the self', *Transactions, Institute of British Geographers*, NS, 18, 1993, 122–139.

Potter, E., 'Modelling the gender politics of science', *Hypatia*, 3 (1), 1988, 19–33.

Pred, A., 'The social becomes the spatial, the spatial becomes the social: enclosure, social change and the becoming of places in Skane', in D. Gregory and J. Urry (eds), *Social Relations and Spatial Structures*, Macmillan, London, 1985, 337–365.

Putnam, T. and Newton, C. (eds), *Household Choices*, Futures Publications, London, 1990.

Queen, S., 'Seventy-five years of American sociology in relation to social work', *American Sociologist*, 16 (1), 1981, 34–37.

Raban, J., *Hunting Mr. Heartbreak*, Collins Harvill, London, 1990.

Rabinow, P., 'Representations are social facts: modernity and post-modernity in anthropology, in J. Clifford and G. Marcus (eds), *Writing Culture*, University of California Press, Berkeley, 1986, 234–261.

Rainwater, L., 'Fear and the house-as-haven in the lower class', *Journal of the American Institute of Planners*, 32 (1), 1966, 23-31.

Ramsay, M., *Downtown Drinkers: The perceptions and fears of the public in a city centre*, Crime Prevention Unit, Paper 19, Home Office, London, 1895.

Residents of Hull House, Chicago, 1895.

Rochberg-Halton, E., 'Object relations, role models and the cultivation of the self, *Environment and Behavior*, 16 (3), 1984, 335–368.

Rokeach, M., *The Open and Closed Mind*, Basic Books, New York, 1960.

Rose, G., *Feminism and Geography*, Polity Press, Cambridge, 1993.

Rose, H., *The Black Ghetto: A spatial behavioral perspective*, McGraw-Hill, New York, 1971.

Rössler, M., 'Applied geography and area research in Nazi society: central place theory and planning, 1933 to 1945', *Environment and Planning D: Society and Space*, 7 (4), 1989, 363–400.

Rudwick, E., 'W. E. B. DuBois as sociologist', in J. E. Blackwell and M. Janowitz (eds), *Black Sociologists: Historical and contemporary perspectives*, Chicago University Press, Chicago, 1974, 25–55.

Sager, L., 'Insular majorities unabated: Warth v. Seddon and City of Eastlake v. Forest City Enterprises, Inc.', *Harvard Law Review*, 91 (7), 1978, 1373–1425.

Sassoon, J., 'Colors, artefacts and ideologies', in P. Gagliardi (ed.), *Symbols and Artefacts: Views of the corporate landscape*, de Gruyter, Berlin, 1990, 169–183.

Sayer, A., 'Realism and space: a reply to Ron Johnston', *Political Geography*, 13 (2), 1994, 107–109.

Schnore, L., 'On the spatial structure of cities in the two Americas', in P. Hauser and L. Schnore (eds), *The Study of Urbanization*, Wiley, New York, 1965, 347–398.

Sebba, R. and Churchman, A., 'Territories and territoriality in the home', *Environment and Behaviour*, 15 (2), 1983, 191–210.

Sennett, R., *The Uses of Disorder*, Penguin, Harmondsworth, 1970.

Shapin, S., 'The politics of observation: cerebral anatomy and social interests in the Edinburgh phrenology disputes', in R. Wallis (ed.), *On the Margins of Science: The social construction of rejected knowledge*, Sociological Review Monograph 27, University of Keele, Keele, 1979, 139–178.

Shields, R., 'Social spatialization and the built environment: the West Edmonton Mall', *Environment and Planning D: Society and Space*, 7, 1989, 147–164.

Shils, E., 'Authoritarianism: "right" and "left"', in R. Christie and M. Jahoda (eds), *Studies in the Scope and Method of the Authoritarian Personality*, Free Press, Glencoe, 1954.

Shils, E., 'Some academics, mainly in Chicago', *American Scholar*, 50, 1980, 179–196.

Sibley, D., *Outsiders in Urban Societies*, Basil Blackwell, Oxford, 1981.

Sibley, D., 'Invisible women? The contribution of the Chicago School of Social Service Administration to urban analysis', *Environment and Planning A*, 22, 1990, 733–745.

Sibley, D., 'Outsiders in society and space', in K. Anderson and F. Gale (eds), *Inventing Places: Studies in cultural geography*, Longman Cheshire, Melbourne, 1992, 107–122.

Sibley, D., 'Gender, science, politics and geographies of the city', *Gender, Place and Culture*, (1995, forthcoming).

Sibley, D. and Lowe, G., 'Domestic space, modes of control and problem behaviour', *Geografiska Annaler*, 74B, 3, 1992, 189–197.

Smart, B., 'The politics of truth and the problem of hegemony', in D. Hoy (ed.), *Foucault: A critical reader*, Basil Blackwell, Oxford, 1986, 157–174.

Smith, D., *The Chicago School: A liberal critique of capitalism*, Macmillan, Basingstoke, 1988.

Smith, M. P., *The City and Social Theory*, Basil Blackwell, Oxford, 1980.

Smith, N., 'History and philosophy of geography: real wars, theory wars', *Progress in Human Geography*, 16 (2), 1992, 257–271.

Smith, S. J., 'Crime and the structure of social relations', *Transactions, Institute of British Geographers*, NS, 9 (4), 1984, 427–442.

Smith, S. J., *The Politics of Race and Residence*, Polity Press, Cambridge, 1989.

Soja, E., *Post-modern Geographies: The reassertion of space in critical theory*, Verso, London, 1989.

Spanos, W., 'Boundary 2 and the polity of interest: humanism, the "center elsewhere", and power', *Boundary 2*, 12 (3), 1984, 173–214.

Spear, A., *Black Chicago: The making of a Negro ghetto, 1890–1920*, Chicago University Press, Chicago, 1967.

Stainton Rogers, R. and Stainton Rogers, W., *Stories of Childhood: Shifting agendas of child concern*, Harvester Wheatsheaf, Hemel Hempstead, 1992.

Stallybrass, P. and White, A., *The Politics and Poetics of Transgression*, Methuen, London, 1986.

Still, J., 'What Foucault fails to acknowledge . . . : feminists and *The History of Sexuality*', *History of the Human Sciences*, 7 (2), 1994, 150–157.

Taylor, P., 'An interpretation of the quantification debate in British geography', *Transactions, Institute of British Geographers*, NS, 1 (2), 1976, 129–142.

Taylor, R. B. and Brower, S., 'Home and near-home territories', in I. Altman and C. M. Werner (eds), *Home Environments*, Plenum Press, New York, 1985, 183–210.

Wallerstein, I., *Historical Capitalism*, Verso, London, 1983.

Wallis, R., (ed.), *On the Margins of Science: The social construction of rejected knowledge*, Sociological Review Monograph 27, University of Keele, Keele, 1979.

Walzer, M., 'The politics of Michel Foucault' in D. Hoy (ed.), *Foucault: A critical reader*, Basil Blackwell, Oxford, 1986, 51–68.

Watney, S., *Policing Desire: Pornography, AIDS and the media*, Methuen, London, 1987.

Williams, P., 'Social relations, residential segregation and the home', in K. Hoggart and E. Kofman (eds), *Politics, Geography and Social Stratification*, Croom Helm, London, 1986, 247–273.

Williams, R., *The Country and the City*, Chatto and Windus, London, 1973.

Willink, H., 'Map-flapping', *Longmans Magazine*, 7, 1885, 404–415.

Winchester, H. and White, P., 'The location of marginalized groups in the inner city', *Environment and Planning D: Society and Space*, 6, 1988, 37–54.

Wolch, J. and Dear, M., *Landscapes of Despair*, Polity Press, Cambridge, 1987.

Wright, E., *Feminism and Psychoanalysis: A critical dictionary*, Basil Blackwell, Oxford, 1992.

Yeates, M., 'Some factors affecting the spatial distribution of Chicago land values, 1910–1960', *Economic Geography*, 41 (1), 1965, 57–70.

Young, I., *Justice and the Politics of Difference*, Princeton University Press, Princeton, N.J., 1990.

INDEX